Jean Allan Owen, Denham Jordan

**Forest tithes, and other Studies from Nature**

Jean Allan Owen, Denham Jordan

**Forest tithes, and other Studies from Nature**

ISBN/EAN: 9783337024550

Printed in Europe, USA, Canada, Australia, Japan

Cover: Foto ©berggeist007 / pixelio.de

More available books at **www.hansebooks.com**

# NATURE STUDIES

# FOREST TITHES

### AND OTHER STUDIES FROM NATURE

BY

## A SON OF THE MARSHES

AUTHOR OF 'WOODLAND, MOOR, AND STREAM' 'ON SURREY HILLS'
'ANNALS OF A FISHING VILLAGE' 'WITHIN AN HOUR
OF LONDON TOWN'

*EDITED BY J. A. OWEN*

LONDON
SMITH, ELDER, & CO., 15 WATERLOO PLACE
1893

# NOTE

*Some of these articles are reprinted from the* 'CORNHILL MAGAZINE,' 'BLACKWOOD'S MAGAZINE,' 'THE TIMES,' *&c., but others are original.*

# CONTENTS

## Errata

THE leaves are falling fast, for a spell of wet weather has been followed by frosty nights, though the days are bright and warm still, and the mellow autumnal tints glow in the sunshine.

At the farm, which is perched on the highest point of the moor, not a sound is to be heard, the menfolk being all at work in the fields. The farmstead is in good order, old though it is, for its stone walls were solidly built, they are thick, and all the timber used was oak, well seasoned. So substantial was the work that, with the exception of a few necessary repairs about some of the outbuildings, all is in nearly the same condition in which it was left by the owners, long dead, who caused it to be built so many generations ago. The land also is to all appearance just what it was when the site of the farm was chosen. What was suited for cultivation was planted then, and the rest left wild, as it still remains, close to the solid path or track that leads from the woods up to the farm ; and halfway up the rush-grown hillside is a bog into which it would be a sorry business to stray. Cattle avoid it by instinct ; no hoofprints are ever seen round that soft place.

Never a moor without its trout stream, or streams, which flow at its foot, cutting sharp runs where, in the course of years, all the earthy matter has been washed away, leaving at last a bottom of clean, sharp sand and bright washed stones.  You can hear the swirl and the splash where it runs through the copse-growth of the moorside, and see the glistening of the pure bright water where it turns into the meadows, as it continually pursues its course from the hills and moors above down to the moors below ; hidden here and there by thorns and brambles, giant 'kexes,' thistles, ferns, and all the other growth common to wild lands.

Rustic bridges, the brickwork of which is as old as the stonework of the farm buildings, carry the cart tracks from one meadow—if land such as this can be dignified by that name—to another.  Where they have fallen into decay and tumbled into the stream many years ago—for mosses cover all the brickwork, over which the water ripples, forming miniature cas-cades—thick planks have been placed, and turf on the top of them.  And as in the course of time the cattle passing and repassing, to say nothing of the carts, have worn the turf covering thin, more has been added, and tough heather tangle worked in with it, until at last a good, solid crust has been formed above the stout, rough oak planks, which is nearly a foot in thickness. As a rule the water is only a few inches deep in the middle; but where it has cut its way under the roots of the trees that line its course from its source up in the hill moors to its final delivery into the river Mole, it is deeper. If you probe with a stick under those roots you will know just where the trout rush to, when they are frightened. At this time of the year they do get frightened, and, more than that, killed; for it is their spawning time—or, as the dwellers on the moor put it, 'the trout are running up'—and the herons are about, taking their moorland tithe, which they will have, try to stop them who will.

For a whole week I have been watching five particular herons busy in most systematic fashion; and also the trout. If you go to work very quietly, you may watch fish, but it is only learned by much practice and experience. A footfall or a shadow will be enough to clear the stream, or at least that part of it which you wish to examine, in a flash.

So we slip down the side of the hedge that leads down to one of the old bridges—they might equally

well be called culverts ; and before reaching it we
crawl on our hands and knees to look through a small
opening there is between the hedge and the brick-
work. The sunshine being warm and bright, the trout
are working on the clean, sharp gravel, winding and
rooting about like eels. Not large fish—they never
are that in the upper parts of these quick hill streams :
the largest of them is only a small herring size, the
very thing for our herons. Scotch firs, of a fine
growth, are numerous along the run of the stream ;
single trees, and again clumps of smaller ones. Some
of the larger ones are sadly torn by the fierce blasts
that sweep over the moors in winter. In the firs,
when the trout run up, one may look for the heron,
one of the keenest and most sagacious of birds. Many
a time has he outwitted and outschemed me in the
days of my youth, and I have always admired and
respected him accordingly. I would fain write a whole
book about the heron alone—percher, as well as
swimmer and wader, is he. From what I have seen
of him in the trees, he is as much at home in them as
is a blackbird or a thrush, and his movements are far
more graceful, notwithstanding his form and size ;
for the bird drops, glides, and walks about the trees as
noiselessly as a grey and white shadow. On the very
topmost shoots of a fir that can afford him a foothold

he will perch, and sit there for twenty minutes, and more, at a time. A few days ago I watched one do this from a place of ambush, which was, like most of those suitable for the purpose, anything but a pleasant one, being a large clump of blackthorns and brambles, in the midst of which I lay hidden. He was not, however, allowed to enjoy his elevated point of survey in peace, for missel-thrushes, blackbirds, and song thrushes, aided by finches, dashed at him, until the long neck and bill went darting and striking out like a silvery snake in all directions as the birds flew at him. Then he was left unmolested, and away they flew to cover, screaming and chirping their loudest.

From my lowly position I had examined some firs that were close to the edge of the stream, thinking herons might be in them, but had failed to discover any. They were there, however, and directly I rose to my feet, almost over my head floated two of these birds, coming out of the very trees I had watched closely, glasses in hand. The heron's positions fall in with chance projections and broken limbs so well—he perches in fir trees by preference—that it is impossible to detect him before he has caught sight of oneself. I put one up recently, and directly in front of me : the bird was standing close to the edge of water over-shadowed by some large alders, the grey mud below being plentifully sprinkled with large open dead shells of the fresh-water mussels. So well did the heron's plumage agree with its environment that, until it moved right before me, I had not seen it. Until then it had only shown a bit of white, like the glitter of the inside portions of the mussel shells that lay gaping on the mud beneath it.

Owls, also, in their plumage mimic and fall in with the tones of their natural surroundings, to a degree past the belief of those who have not patiently watched them in their own haunts.

And often the most patient and constant watching is unattended with success. Early and late had I been abroad of late, and had come home quite disheartened with failure, and then I succeeded in my quest owing to my coming unexpectedly on a large, partially drained moorland fish-pond, whilst in search of an artist friend of mine who, as I had been told, was painting in that direction. After stumbling on bits that were enough to drive a would-be artist wild, through my field-glass I discovered him settled right out in the most treacherous part of the moor. Not without much cautious travelling did I reach him. Whatever possessed him to go and paint in one of the worst moor swamps of the district? I asked. 'To get a faithful study of the moor mosses, sulphur yellow, and rich dark brown, with the rush clumps scattered about over it,' was his reply. This was his foreground ; in the distance he had two cottages, and beyond were the fir woods.

On his feet were huge wooden sabots, lashed tightly to his leather boots. These, he informed me, were made by himself—sepulchral-looking articles they were. He had placed four coats of black paint on them, he said, both inside and out. Anyway, they answered his purpose admirably, for although the water had been flowing under his feet beneath the moss all the day, no damp had reached him ; and the result on his canvas was all that could be desired. To my request that he would favour me with a dance in those coffin-like damp-protectors before I left him, he replied that it would hardly be wise to dance about much, for only a few days previously a couple of cows had got bogged just below the very spot on which we were, and had the farmer's men not pulled them out with ropes, they must have been smothered.

To return to my moorland fish-pond. The sight of five herons sailing in wide circles over any spot would be sufficient to attract the notice of any naturalist. The birds were shouting hoarsely to each other, after the fashion of rooks, but far more noisily, just out of gunshot, but within capital distance for a rook rifle. My appearance in the open did not cause the slightest difference in their movements : they kept passing over me in their wide circles, which was quite contrary to their usual cautious method of movement. On diving into the deep hollow of the moor I discovered the cause of this. A sluice at the further end of the pond had rotted and given way, letting great quantities of fish down into the moor stream, to the great joy of all the prowlers in the district, both furred and feathered, particularly of the herons. And the foresters, noting the birds at work in the stream, profited by the sight, and had fine fish also, for the mere picking of them up out of the water, which was too shallow, now that the stream had gone down, to cover them.

When the herons found that the stream was not safe for them, they confined their fishing to the centre of the water that remained in the large pond, where they were in no danger, and where their finny prey seemed literally ready to jump into their mouths ; for the large fish, pike, perch, trout, and eels, all of them that remained there, chased the smaller fry continually. So eager were they, that the shallow edges were all on the ripple with small fish that had rushed thither, where those that fed on them could not follow them. But even this did not completely frustrate the

murderous intentions of the bigger fish, for some of
the pike rushed on to the mud and perished there,
being unable to get out of the ooze. This accident
had brought the herons to the spot in force. The
immediate reason of their hoarse cries and their
circling flights at the time I came up was soon explained. Two men were repairing the broken sluice-
gate at the further end of the pond, and although the
birds might have settled and fed to their hearts' content, they would not do it whilst the workmen were
there.

The sluice-pond job is finished, but the pond is
still very low : it will take a long time to fill up the
large space again, even though it is supplied by a
trout stream. So the herons—more have come since
the first ones prospected and returned a favourable
report in their own bird fashion—have it all their
own way there. They range the edges of the shallow
pools as well as the main current in the centre, and
when satisfied to the full, they rest in the trees near,
where, I am happy to say, they have not been fired
at. A few gentlemen who rent these moorland
districts for the rough shooting they can get in them,
have given instructions that the herons shall be left
undisturbed—a very wise precaution, for the diet of
these birds is a varied one, and some of the creatures

that have their habitat where the heron hunts for his living are far too numerous.

As long as the moors remain there never will be any lack of trout in the streams, yet even those who have the right to fish there do not always get them. After all, a shilling's worth of fresh herrings would, in my opinion, surpass even a fortunate catch of such small fish, and it seems a pity that so many of the little spotted beauties should have to go to make an ordinary ' fry,' as they term it. However, there is no perceptible falling off; others come to take the places of those that get caught ; and, happily, many who have the privilege of fishing not only do not use it much themselves, but they take very good care that some who have not that right, yet could very well dispose of the fish, or feed themselves and their children with it, do not get the chance. When the moon shines, our herons do not let much escape their sharp eyes and bills.

Calm and beautiful as the moor farms look in the summer and autumn, surrounded also by the loveliest woodland scenery in England, in the long winter time they look the picture of dreariness ; the snow frequently drifts up about them so as to shut them out from the rest of the world. In the lanes it is shoulder high where it has been blown by the fierce gusts of howling winds that sweep over the uplands. A fortnight of such weather as this brings the wild creatures close up to human dwellings, and particularly to these lonely farmsteads. When the light shows red through the window casements, before the stout shutters are pinned up for the night, it is the signal for the fox to come and see if he can pick up a supper somehow. It will be through no fault of his if he fails in this, if there is anything about that he can kill and eat, from a turkey down to a poor starved blackbird, or from a hare to a mouse. I have seen some grey dog foxes that I should not like to be shut up in a small room with, if I had nothing but a stick in my hands. A hard fighter is Reynard, when cornered ; he snaps like a wolf and is active as a cat. He makes short work of anything he can capture. Domestic fowls have, some of them, an aggravating way of preferring to roost, like pheasants, in the trees at night, and to make their nests away out in the moorland or under the trailing plants and bushes. That jubilant cackle, however, which the successful hen cannot control herself sufficiently to repress, after she has had the cunning to hide away in her nesting place, betrays her to other creatures besides her lawful owners. She may even have been prudent enough to wander a little way from her eggs before she commits herself to it; but the fox knows the meaning of the cry, and profits by it. Although, as a rule, he prowls about in the night time, when his cubs begin to eat flesh and are hungry he is not particular as to the hour, provided the locality be a lonely one. I have known the hen's jubilant voice stopped abruptly, and on going to the spot whence it had sounded have found a few feathers and the marks of a hurried scuffle, but no fowl.

During one bitter winter I was, in the pursuit of my calling, located in the heart of the moorlands, where my home for the time was in the cottage of a hearty old couple. After supper the three of us were wont to occupy the old-fashioned chimney settles— 'father' and 'mother' on one side of the fireplace, and I on the other. I happened to say that I had overheard some men complaining that they had tracked a fox, and found a couple of their fowls buried.

'Ah!' said the old man, as he puffed slowly at his pipe, and patted the red-hot ash with his finger, 'I killed a fox for that werry self-same thing; but,' he added almost in a whisper, 'nobody knowed on it. You 'members it, mother?'

'Massy, oh!' rejoined she, 'I wouldn't hev sich

another set out as thet 'ere fur all the spring chicken
as iver wus hatched. I wus feared the Squire wud
ha' got hold on't. A good master he wus, but he
showed no marcy to any thet hed killed a fox on
his grounds, an' they run middlin' wide.'

' Mother kep' her fowls for market,' resumed the
old man—' they paid middlin', most special in airly
spring time ; 'twus jest sich a winter as this 'ere, an'
some o' them fowls went. She wus 'mazin' vexed,
for they kep' on goin'. I knowed 'twus a fox, but
didn't let on to mother, on'y one mornin' she seed
him cum slippin' roun' on the snow. Thet did it.
When I come home to supper that same night she
looks me most oncommon straight in the face, an'
she says, " Kill that 'ere *thing*, father, or 'twill ruin us."
She wus middlin' spirity that time o' day, I ken tell
ye, an' I knowed she meant what she said. Well, to
cut it short, I got one o' mother's oldest hens, an' fixed
her up so as the fox could see her, in an empy pig-
stye. Right in front o' the old hen, about a foot
away, wus a trap—a real good un. We went to
bed, and in the middle o' the night we wus 'woke
by the most desprit row. " Father," sez mother,
" you've ketched thet 'ere varmint o' a fox. Git
up, an' take a prong wi' yer." (Prong is local for
*pitchfork*.)

' Up I gets, lights my lantern, takes the prong, an'
goes out. Massy, oh ! the old hen wus hollerin' like
mad, an' the fox wus bangin' about in thet 'ere trap—
the way he flew about, an' bit, wus a sight to see. I
settled un with the prong, an' the old hen too, to stop
her hollerin' ; an' I buried the pair on 'em, in one
hole, a goodish bit away from the house ; but 'twus
a long time afore we felt easy like ; fur 'twas a
most 'menjous row to be heerd in the dead o' the
night : ef anybody hed bin about they must ha'
heerd it.'

If very stringent measures had not been taken for
his preservation, the fox would have long ago shared
the fate of others of our wild creatures more vir-
tuous but less fortunate than himself. Foxhunters
to the manor born have, of course, their own views on
the matter, and these are very properly universally
respected. Besides this, it is not pleasant for some
others belonging to a certain class who have
struggled up into Society—with a big S—and wish
to keep there, to be cut dead because they have
given their keepers orders to kill foxes. So Master
Reynard goes on taking his tithes, especially when
he has got cubs, for whom it is his duty to pro-
vide. And his foragings round the lonely moor
farms are not by any means a dead loss to those

who suffer from his depredations. If complaint is
made in the right quarters, compensation is generally most cheerfully given. Yet I have known
some unhappy souls grunt even after they have
received more pay than the case demanded. It
is a hard thing to please folks. A fox, dog or
vixen, never comes near man, woman, or child if
he can help it ; but if hemmed in, he would be as
dangerous as a collie dog is under the same circumstances.

At the foot of one of the South Down hills, where
there is a long strip of coarse grass and moss, a
quarter of a yard wide and over a mile long, there is one
of the abiding places of the badger—one well known
to me. The land was once broken up by the plough,
but this proved useless labour and expense, and now
it is a paradise for wild creatures. That prince of
British butterflies, the purple emperor, floats and
dashes over this bit of moorland valley, crossing from
one belt of oaks to another. The mossy-bee and the
humble-bee have their homes in the moss surface, and
our old friend, in his grey, black, and white coat
scratches them out with his claws, or roots them up
with his nose like a pig. In some of the moor
districts he has been wiped out, and much it is to
be regretted that those who have had the power to

prevent this had not also the knowledge and the disposition to do it.    One by one our wild animals and birds are dwindling down as to numbers : not shot or trapped on account of any real harm they may do, but for the money the creatures fetch, dead or living. One hears in all directions of the harm done by rabbits, although the Ground Game Act has been passed.    What can be expected when the creatures that were there to keep the pin-wire vermin in check are killed ?    Besides which, these are bred in vast numbers expressly for the markets.

Our vermin-killers—polecats, stoats, and weasels— have been captured and bought in wholesale numbers, in order that they may be exported to some of our colonies to kill the rabbits which were formerly introduced there by mischievous bunglers, and which now prove a curse to the land.    Change of climate and habitat alters the habits of creatures that are not indigenous.    It will be found that polecats, stoats, and weasels will turn out another curse ; instead of killing off the rabbits, they will certainly prove an inveterate foe to poultry and to the small animals that are native to the soil.    I speak feelingly on this subject, as one whose opinion on the matter has been often asked and freely given, yet never followed.    This acclimatising question is a most unsettled one, but it will have to be studied, to spare much future trouble to us and our colonies. Soon we shall be cursed with a plague of rats and mice that will not easily be kept under.

Man considers himself the lord of creation, as well as of the soil he buys or inherits, but some of the changes he brings about are to my mind matter for great regret. Quite recently I have seen pheasants take the place of blackgame in one district I know. Only a few of the latter lingered there, the remnant of a once fairly numerous family, real natives of the moors, not imported birds ; but they are gone now. Where they once fed about the rush clumps, on the rush seeds, a keeper's cottage now stands, with dog-kennels attached. Instead of one flushing a black-cock, one sees a pheasant spring up, and the croon and play of the former has given place to the drum and challenge of the latter. And two years ago I saw preparations made on a secluded side of the moor to turn the locality, where only wild ducks, woodcocks, snipe, and blackgame used to have their habitat, into a cover for pheasants. The job is now completed, and it will answer its purpose to perfection, but the alterations have certainly taken away some of the beauties of that wild hillside.

To return, however, to our moorland tithe-takers.

C

The polecat, who used to be one of the chief of these, is now almost a thing of the past, and when he is seen or captured the circumstance is made a note of. Like the rest of his species, this very powerful, and in his own domain most useful, little creature carries his prey like a retriever. From the polecat to the weasel, the strength of the family is something remarkable. It is a wonderful sight to see one of them come bounding along, holding a prey as large and as heavy as itself off the ground by the middle of the back. A few days ago, a friend of mine—a keen observer of wild life, the result of whose observations will, I trust, ere long come before the public— saw what he took to be a lump of dry hedgerow plums blown up by the wind over a green ride towards him. As it came nearér, it proved to be some creature bounding along bearing a half-grown rat, and, when within a yard or so of the spot where he stood, the rat was dropped on the turf, and a weasel looked at him as only a weasel can look. The animal was only aware of the proximity of a human when he got wind of my friend; in fact, the rat he carried before him prevented the weasel from seeing what was directly in front of him. The man lifted the rat, noted the way in which it had been killed, and then placed it where the weasel had dropped it, so that the latter might fetch it after he had gone.

Rats do not, of course, frequent waste lands, since there would be nothing for them to feed on there; but most of the cottagers who live on commons keep pigs and fowls. From one of these homesteads the rat was being carried to a wood-stack, where it could be consumed in peace. Determined foes to rats and mice are the weasels, also to frogs and some other small deer, and for doing man this good service he kills them or exports them where they are not wanted. It is true that stoats and weasels kill rabbits, but for one rabbit they will kill forty rats and mice. I only wish the whole of the family were more numerous! Five years ago we had a great plague of mice in the woodland districts. The pests invaded the gardens of those who had large houses near the woods, cut down the flowers in the borders and dragged them into their holes, and nibbled the wall-fruit. In fact, the folks had to cast about to find measures for their destruction. When things were at their worst, weasels made their appearance—whole families of weasels, just as the short-eared owls suddenly appeared during the more recent plague of voles, on the southern up-lands of Scotland, in large numbers. These owls have also remained and bred in the district, to the great relief of the land. It is to be hoped they will be unmolested and their usefulness universally recognised. Sir Herbert Maxwell has used his local influence, I believe, to this end.

The weasels were noticed all making their way to the parts where the mice were gathered. Then the mice shifted their quarters, but the weasels followed. Two or three families, the old ones and their half-grown kittens, will soon move mice where they are about. Four or five ferocious old grey rats will kill more poultry—of all kinds—and steal more eggs than all the weasel family in a district.

Anyone walking on the roads that are only separated from the cornfields by low hedges in our outlying country districts, in the dusk of the evening, after the corn has been carried, can see some very pretty hunting; for then the rats come to the fields and live for a time in the hedgerows. If you go very quietly and look over one of the field-gates, you will see dusky, bunched-up forms some eight or ten yards from the hedge; rats these are, feeding on the scattered grain. Presently they go loping up and down, with their peculiar gait, making for the cover of the hedge. There was nothing apparently in the field to have alarmed them, so we look upwards, and see at once what it was that caused them to seek the cover of that hedge bottom. Two brown owls are beating over the field in quest of rats and mice, no matter which ; the first that can be got at lives but a very few moments after it is gripped and bitten through the back of the neck. Their hunt, this beat, has been useless ; they flit over the adjacent fir-trees and for a time become invisible to us. One by one, by threes and twos, out come the rats again ; the owls are not overhead, so some venture out a longer distance from the hedge. Presently we hear a rare scamper and some squeaking, for the owls have changed their tactics. They have come down the side of the hedge this time, and flapped again into the fir-woods, each with a rat in his grip. A very fortunate matter it is for the owls that no one in a velveteen jacket and carrying a double-barrelled gun is at hand, or their spread-eagled forms might have ornamented the gable end of a dog-kennel next morning. One of these gentlemen, to whom I lately appealed for their protection, replied, ' I ain't *sin* 'em do what some says they will do ; but I kills 'em when I has the chance, to keep 'em from doin' of it.'

When will things be set right in these directions ? I often wonder. I have seen the inside of wheat-stacks eaten out, befouled, and ruined by rats and mice ; and still owls are killed and farm-labourers are rewarded

for killing stoats and weasels.  A faggot-stack, also,
I have known half pulled to pieces to get at a poor
little mouse-killing weasel that had taken refuge there
from its pursuers.

Even that glorious insect-destroyer, the great green
dragon-fly, that helps to clear country lanes of such
winged pests as hornets, wasps, and the ferocious
stoat-flies—'stouts'—has evil properties attributed to
it by the rustics.   I have seen country children—and,
indeed, grown-up people—show far more fear at the
sight of that swift-winged beauty than they would at
hornets or a wasps' nest.  'Them 'ere things is adder-
spears, an' it ain't safe tu meddle wi' 'em.'  This superstitious belief, in fact, saves the grand insect from
being killed.   It is a very rare thing to see a dragon-
fly captured in country places.  The farmers' lads
also call the great dragon-fly the hoss-stinger.  Just
a word of advice to those who have delicate fingers :
this fine fly, if held incautiously, can and will bite
pretty severely.  So will the very large and handsome
garden spiders which weave their wonderfully geo-
metrical webs, suspended from twigs and lashed to
branches, in the latter part of the autumn.  A colony
of very fine ones have located themselves in some
box-trees close to my back door.  The number of
insects they capture in their nets is something wonder-

ful, particularly large bluebottle flies. They do not get all they capture, though, for wasps come and take the bluebottles out of the web piecemeal—the head goes first, and the body is left for a second journey. Full well do the spiders know the difference in the sound of the hum of a fly's wing and that of a wasp. When the latter comes to rob their webs they do not show themselves ; but directly a bluebottle gets meshed, they make a tiger-like rush for it.

The hedgehog, urchin or hedge-pig of Shakespeare, is very little understood so far as his habits are concerned. I have seen some recent statements in various papers to the effect that hedgehogs will kill young fowls, as if this were a new fact for the edification of their readers. Not only will they kill the young ones, but the old ones are not safe from them, if in a coop. The foot of Master Hedgehog is sure to be put in or on it if he gets the chance. I have even seen our friend described as ' our poor little persecuted English porcupine.' That he can never be, for his habits are nocturnal, and no one considers him worth looking after, unless it be a gipsy, who has a taste for baked hedge-pig. He is not so harmless, however, as some other creatures. A very large hedgehog is quite capable of killing a wild rabbit. He can depress his spine and squeeze through places one would have supposed too narrow for him.

The distance this nimble creature can travel over is quite remarkable. Out in his own domain, the woods and the fields, he is perfectly harmless. Some hedgerows are frequented by numbers of them. There they lie curled up among the dead leaves in the daytime. In other hedges you may search for them in vain. All would go well if the bristly little animal would keep his own place. But he will not do this. He sniffs until he finds something to his liking, to which he applies himself with pig-like determination. Sometimes this happens to be a trap baited with portions of rabbit's flesh, where he gets caught, and whines most pitifully.

But sometimes, when the farmer's wife goes to look at the coop out on the grass in front of the house, she finds her chicks killed and eaten, and their mother bitten and nearly dead with fright, the sole cause of the mischief being a great hedgehog, which is trying to get out of the coop: not so easy a matter now that his stomach is full of chicken as it was for him to creep through when empty. In such a case he is usually promptly settled by the points of a bright hayfork.

But, after all, when the matter is fairly considered

without prejudice, the amount of tithe taken by the
wild creatures of the moorlands is very small. The
creatures that are able to do much mischief are now
very few in numbers. As to the birds, the moors are
not their favourite dwelling-places—at least, not of
our common birds, which keep to the line of cultivated
lands. You will see trees laden with fruit in the
cottage gardens, placed in spots few and far between
on the moor; cottage and garden in its own small
enclosure. The jays may squawk in the cover round
about these, but they let the fruit alone, for they would
have to cross a considerable open space from any given
quarter to reach the fruit. On the whole, considering
all the devices employed by man to circumvent these
tithe-takers, it shows great sagacity on their part
that they are able to get as much from him or his as
they do.

# EYES AND NO EYES

As a child I had two great desires, and I do not remember indulging particularly in any other : one was that when I grew up I might paint pictures of the wild things that surrounded my lonely home, the other that I might have money enough to buy books about them. I am thankful to say that both these longings have been in a great measure gratified. In my wildest day-dreams, however, I never aspired to writing myself about the creatures. That has come about since my hair turned grey and my hot blood has cooled down a little. I am sure of one thing—that a man who lives amongst the so-called working-classes, and who has also unrestricted intercourse with foresters and the more intelligent rustics, has opportunities for gaining a real insight into wild life such as many a student of nature, who may have been what is from the world's point of view more fortunate in the circumstances of his life, cannot have.

Said one to me lately, ' I have read your books. But do you really see all those things when you walk about ? '

' Not in the streets of Dorking town,' I replied.

' Because really, my dear sir, do you know that in all the time that I have been in this neighbourhood I have never seen a tithe of what you have written about.'

The old story of 'Eyes and No Eyes,' I said to myself ; also that if some folks had eyes at the back of their heads, as well as in front, they might wander far and see little.

Another man observed that, although he had ridden in a cart for many years in all directions along our high roads, he had never come across any of the creatures I had written about. As the vehicle he used made nearly as much noise in its progress as a goods train might do, it was hardly to be expected that he would.

Readers can easily see that my range is not a wide one—it is only the common objects of the hills, dales, and waters of a limited area that I describe ; yet some pains have been necessary even to do that, and in the pursuit of the rail family alone I have often supped sorrow ; while to verify a fact or two has cost me week after week of hard tramping

many a time. In this way the naturalist fits himself
for writing just a small portion of a bird's life.
Sometimes the long twenty or thirty miles' walk has
been to no purpose, and, after giving up the pursuit
as a bad job, I have turned my face homewards, and
then found the very bird I had gone so far in search
of within a twenty minutes' walk of my own door.
By patient watchings and waitings, on the part of
many different naturalists, fact has been added to
fact, until the whole life of a creature, furred or
feathered, has been placed before the public, in order
that those whose labours confine them to crowded
centres of industry, but who have strong sympathy
with life in the open and the creatures who are able
to enjoy it, may understand what are the real lives of
the animals and birds.

It is a difficult matter to please some of these
would-be students, however, and a short time ago an
amusing scene was reported to me. A gentleman,
who had read a certain article which had been written
by myself, came many miles to see a woodland river
and an old weir, the haunt of the otter, which I had
there described. When he arrived, however, a band
of workmen had, unfortunately, just finished building
a new weir, and they had also cut down all the
alders, willows, sedges, and other growth along the

river-side, leaving just the stems. This is done every five or six years, as the case may need. Unfortunately, he arrived at the place just as the two jobs had been completed. Hither and thither he rushed, until someone asked him if he had lost anything, or if he were looking for someone. Then he gave full vent to his injured feelings ; he said, in fact, that he had been swindled—that the writer had drawn very largely on a very fertile imagination. This was perfectly unintelligible to men who knew nothing about such a book having been written. One looked at the other, and then touched his forehead, muttering 'Balmy.' This the gentleman heard, and naturally resented.

'I have been grossly imposed upon ; some people's geese are swans. Where is the old weir ? ' he asked.

' Just been pulled down and a new un built. Don't ye see it ? '

'Why has it been pulled down ? '

' Well, if you ask us, they pulled the old un down to build the new un. Any fool can see that.'

' But what a complete swindle—there are no trees here ! '

' Maybe you wants to buy the timber and the faggots. If ye do, ye needn't put yerself in a tantrum ; none of it ain't sold yet.'

Unheeding this, the gentleman went on complaining bitterly ; as to owls, he didn't believe there was one of them about the place ; and talk about the yikeing laugh of the yaffle, it was sheer humbug.

'What's that you say—no yaffles ? ' replied one of the men, much amused. 'They're nearly as common as blackbirds in this 'ere park ; and owls too, if it didn't happen to be just mid-day. If you was to see them yaffles, and to ask them to holler, I dessay as they'd do it for ye. But do ye know as you're trespassing on this 'ere land ? '

'I have come to see some otters.'

'Then you've come on a fool's errand. Did ye think they run about here like sheep ? I can get ye one for a suvrin. Ye don't want one ? Then you walk off here ; for I tell ye, ye are trespassing.'

It was certainly very disappointing for our friend, but rather hard lines that I should be held responsible for his disappointment.

I have received many most kind and sympathetic letters from ardent lovers of the birds, ladies as well as gentlemen, asking me if I would give them the exact localities where I have seen such and such migrants drop in their flight ; or where such and such a sequestered pool happens to be situated. Some of these I would fain answer; but I make it a rule not to betray the whereabouts of any of the wild creatures whose secrets I have surprised.

As I said before, it is really hard work watching any members of the rail family; for the cunning of these birds exceeds all belief, and the places they frequent are nothing but quakes. Early in the morning and late in the evening are the times for getting a glimpse of them—that is, if you are lucky in this, for they move about more like rats than birds. Then the midges rise in clouds and sting you most horribly, swamp lands being the abiding place of these insect pests. They form a portion of the daily food of the rails in their nesting time.

And perhaps when you have even offered a rustic, on whose plot of land you want to trespass when in quest of your bird, a sum of money for a small grey and brown bird that you have seen, if he will snare it for you or allow you to snare it—some of the garden plots are only separated from the swamps by a turf wall—he wonders, and some little diplomacy is necessary; for the owners are very tenacious about right of way, and will resent any attempt at trespass, even fiercely.

Many have been my failures compared to my successes in hunting after wild things. Let me

describe the haunt of that spotted crake, or, as it is
usually called, spotted rail, which, after all, did *not*
come into my hands.   A water rail is a bad enough
bird to look up ; a spotted rail or crake is worse.
Both can run, climb, swim, dive to perfection.   Even
in thick tangle their movements are as cautious
as when they, with coots and moorhens, visit the
gardens on the edge of the swamp from which these
have been reclaimed.   This they will do in order to
pick the hearts out of the ground crops that have
just been planted.   How they find them out we do not
know ; and the mischief is done late in the evening
and at night.   I have seen them at it many a time,
and when the pools or ponds are close to cottages
situated on estates this has to be put up with : no
shot may be fired, or noose set, where the gentleman
in the velveteen suit walks round.   I do not intend
to enter into this matter : not being my business, I
leave it.

A tramp through meadows, with the Mole
twisting in the most erratic manner, for three miles,
brings us to a clear pool of considerable extent,
fringed on one side with dense aquatic tangle.   Hay-
ing time is over, so we go right up to the bank of
tangle from the meadows through which the river
snakes.   As sedges, bullrushes, stunted willows,

meadow-sweet, alders, and loose strife do not enter into the composition of hay, the mowers have simply cut as far as the tangle, leaving that as it was. I know what I am about to suffer before I enter it; for I have left off smoking, and the day is what the people in and about the fields call a dead-hot day. Not a breath of air is stirring and the 'stouts' bite most ferociously. At this time birds and fish revel in insect food; it drops down on the water or skims over the surface, and the fish lazily gulp it down; in fact they are gorged with it.

I have entered the tangle at the thinnest part, for I wish once more to confirm something that I have stated elsewhere, namely, the pike watching for young water birds. If I could see that every day I should not tire of watching it, for the birds know of their danger, and guard against it as well as they can. Not many yards I have crept along, parting the tangle gently to right and left, without making a rustle, getting the backs of my hands covered with midges, whose bites I bear somehow; then I drop down and crawl to the edge, looking through a rush tuft which I part for the purpose, and not a yard from me I see a pike—not a large one, for no large fish are here, they do not thrive in this pure spring water. From two to three pounds weight will do, but nothing over

D

that ; they are hungry, however, as wolves. There he lies, motionless, just behind some withy roots that run down into the water. Presently there is a ripple on the other side, then another ; the fish's fins begin to quiver, but his tail does not move. That quiver of the fins draws him nearer. But there is only a water vole with a bit of sedge in his mouth swimming along, and not the least notice does the fish take of him. Although at other times voles would come to grief, they are safe when young water birds are about.

If I thought fit to go on looking for pike, I should find one watching every few yards. One side of the pool is bordered by open meadows without any tangle. At this season, although very numerous in the water, the pike do not frequent the open ; they will come back on that side presently, when no more birds are to be had. From a small puddle some bird rushes, and then flutters broken winged. I know what it means —the bird has young. It is all a sham, but a first-class sham it is ; another step, and away go a clutch of young moor-hens, like a lot of black mice on stilts. Willow wrens chide me—I could almost touch some of them—but I press quietly on, smothered by dead tangle and midges, though it will be impossible for me to put up with much more of it. Was that a rat that glided through the bunch of king-cups ? No, it shows for one moment, flirting up its tail; it is a spotted crake, the very bird I had tried in vain to get.

Cold spring water was freely used to my face and hands for some time after leaving that pool-side tangle. Birds do not very often come to those who are looking for them; you have to go to them, if you can get there.

It is winter, hard, bitter winter; the snow covers the shingle on the beach above high water-mark; the sky is lead colour, and the water looks like ink, broken at times by the spiteful spit of the crests of the waves, that can barely lift, because the wind is blowing down on them, right on shore. When the waves break, they run up with a hissing noise that tells plainly what is coming.

They rush like huge snakes up to the snow-line on the beach, sending blinding salt spray in all directions; and rush back again, the stones rattling on the beach as though an express were at full speed there. It is a gale already; it will be bad presently; these signs are only preliminary ones, for the waters from the North Sea will come roaring in, a little later, up to the sand dunes. A dreary look-out there is—a long stretch as far as the eye can reach of snow, dark sky, and darker water. In the foreground are a few blocks

of rock, about the height of a man, and behind one of
these a fowler is crouching.  That great mass of white,
about a mile out, is where the wild rush of waters is
breaking up in foam on the sand bar.  You can hear
the howl of it ; not a goose, duck, diver, or wader will
be able to rest there, before this gale is over.

Fowls know what is coming ; it is from them, of
course, that coast people gain their knowledge, in a
great measure, about atmospheric changes.  At one
time, well within my recollection, the fowls' movements were the only signs they had to go by.  Bad
as the weather is, fowls are now eagerly feeding on
the very edge of the tide—curlews these are.  Something moves a whole mob of them, and they rush up
shrieking as curlews only can shriek—just over the
edge of that hissing surf.  Ears and eyes are on the
alert ; and as they pass that rock the fowler fires, and
eight birds drop in the water.  Without one moment's
hesitation the gun is dropped in the snow, and he
dashes in up to his middle in water ; five he gathers
and brings on shore, then he thinks he will strip and
swim for the others ; but luckily he is restrained, and
the other three birds drift off.  I have done foolish
things like this in my time, but this was done by a
brother naturalist, who has just left us.  I admired
his pluck, but, being twice his age, I at the same time

pointed out that the thing might be done once too often.

In search of birds that frequent salt water and the shores, I have never failed to find some of the various species ; quite sufficient indeed to satisfy me. The plan of proceeding is a very simple one ; certain birds at various seasons—spring, summer, autumn, and winter —frequent particular localities. By knowing these, and going direct to the places, you will find your bird or birds, and no valuable time will be lost in searching. Fashionable watering-places, however, had better have a wide berth given them.

The attitudes that some large birds place themselves in would protect them in comparatively open places. Unless one had seen it, one would hardly give the raptores credit for this ; but they practise it to perfection. This class of birds has been my favourite study from boyhood, and I shall never be able to finish the study, for something fresh is continually coming before me.

The eagle, for one, places himself in strange positions, a mere bundled up bunch of feathers to look at. A friend of mine who recently visited Achill Island, the island of the eagle, was on the cliffs with his wife and some other friends, when the lady saw in a cleft of the rocks close to her what she thought was a splendid tuft of feathers that had been blown there by the wind. On stooping to pick them up, out from the cleft dashed a magnificent eagle, leaving behind him a tuft of feathers as a memento.

Vigilant as the bird is, he is frequently walked over. After gorging he gets drowsy ; indeed, it is only under these circumstances that such a thing could take place.

Falcons, hawks—the larger species—can compress their feathers and look very slim, if they think it necessary to do so. As to the owls, they can hump up into any position they think most suitable. It is useless to look for these self-preserving traits in any of the family kept in zoological collections, for the birds are so accustomed to see large numbers of people passing and re-passing, or standing in front of them, that they treat the whole matter with perfect in-difference. They know that at a certain time their food will be brought them, and that they are other-wise perfectly safe. Then the raptores in a wild state have a bloom on their plumage, like the bloom on a bunch of grapes, which is not often seen when in captivity.

There is a sameness to be seen in the habits and haunts of all shore birds, no matter where they may be. All shore-shooters know this and compare notes about it. Flats of slub ooze are the same in one locality as another, and beach is beach. As to the sand dunes, or sand hills, they are all alike in general features, go where you will; the only difference is that some are of very great extent, and some comparatively speaking small. One instance out of many that I have witnessed will prove how completely a bird's plumage may mimic the surroundings that it lives and nests in.

We are standing between broken hills of the sand dunes; great humps there are, that will topple over during some high tide, for the swirl of the waters has washed the worn bases of several nearly through. In between, large patches of purple-grey shingles have been washed. I have called them purple-grey, as this is the general tone of them; but broken shells, white pebbles, and the thousand and one atoms of flotsam and jetsam that litter the foreshore, make a glitter that is very deceptive, if you wish to pick matters out in detail.

Tufts of marum, or bents—for this creeping, wiry, grass-like growth is called by either name—hold on where there is the least chance; if I were asked to give this a name, I should call it sea couch—for it will couch and hold on anywhere. A most valuable ally it is where it is found; it binds the sand dunes by its network of creeping roots. This, with sea holly, sea thistles, sand convolvuli, and a few tufts of blite, is all the vegetation that grows here.

Pools of water left in some hollows by the last storm glitter in the sun ; so clear is this very salt water that a pin could be seen at the bottom of the deepest pool. If you turn your eyes inland there are the rabbit links, and as we are in the hollows that is all we can see in one direction. Looking seawards we see a vast flat of sand, for the tide is out. A solitary gull flaps overhead, the only form of life that shows itself. It has a beauty, although a weird one, this great sand-flat with the dunes as a border to it ; but I should not care to spend a day here alone. Others have visited the place and made the same remark, for the strange shapes into which the water had formed the sand hills, some of them fallen, and others ready to fall, used to give me the impression that a small part of our island had been utterly wrecked.

The ringed dotterels nest here ; as we tramp over the shingle two rise right in front of us and settle on a sandy knoll near ; we can just make them out, and that is all. Then they pitch on some beach a few yards away, and we do not see them again ; all our creeping, hiding, and waiting will not benefit us one bit. Still we do not like to be beaten if we can help it, and we try again, but all to no purpose ; then, just as we are leaving, the birds flit up on one side of us, and we are baffled.

If any one of our readers will visit the New Natural History Museum at South Kensington, in the fine room filled with splendid cases of British birds he will find a case of dotterels and their young ·on some beach shingle. They are perfect in their setting up, even in the most minute details. When looking at these he will see at the first glance what I have endeavoured to explain here ; and that is why ·dotterels and their young are invisible on shingle.

## UNDER GREEN LEAVES

'YOU like rough ground,' said one of my friends one day, 'so I think you would find something to suit you in the Long Valley Moor. You can go just where you please, and if you like to catch a brace of trout to paint from, you are quite welcome to them.'

Thick warm mists shroud all things as I leave my home for a six miles' walk, wishing to see the said valley lit up by the morning sun. Only those who have walked through the woodlands in all their fresh tender greenery, when the dewdrops are glistening like diamonds on the foliage, and the grasses nod over the narrow track, bowed down by the clear beads of moisture that cling to their edges, can form any idea of the beauty, the calmness, and the good knowledge that are to be found under green leaves.

The rough weather has gone, winter shifts are a thing of the past, and the birds sing, for it is their nesting time. Man himself, in spite of all his cares, brightens up and looks for better times. The Combe farms that we pass on our way are still quiet ; with the exception of the carter and his mate, who are going to the stables, no one is moving. Not a rooster crows ; it may be that the birds know that the fox is about : birds do know when it will suit their purpose to keep quiet. How often have I wished that there was a chance of a fox coming near our house just to frighten those many roosters into silence that, through moonlight nights, will crow by turns almost without intermission. It is all very well for poets to write about 'Chanticleer, bright herald of the morning.' This lively fowl has been rather a favourite with our verse makers, especially the more moral of them. I fancy they have been mostly familiar with the common barn-door bird, which simply crows ; they cannot have suffered from the roaring of the Cochin China fowl, which after so-called judicious crossings—a fatal blunder—has become such a favourite on the outskirts of country towns. Owls never make night hideous as do these giant roosters ; I have found two or three of them, in concert with a donkey that grazes near my dwelling, more than a poor tired naturalist could bear after a long day's work.

But here all is quiet ; the creatures that were at large during the night have gone home, or most of them. Two downy forms float by me and shoot under

the thatch of the barn.  These are brown owls that
have come up from the meadows.  They have not
yet finished their hunting, and they leave the barn to
float down to the grass again.  So much the better
for the farmer ; if he had twenty owls instead of a
couple about his premises it would be a good visita-
tion for him, for when the grass is mown, a boy will
be put to walk behind the mowing machine, where he
will be sure to be accompanied by another, a volunteer,
who goes for the sport of the thing.  It is a much
envied post ; the first lad is the mousekiller, and he
is armed with a stick three feet or so in length, of
ground ash by choice.  The invited friend must cut
his from the willows.  How often have I watched
this proceeding.  The boys will wait for the machine
to make a couple of swathes, one up and one down,
then they follow it up on each side.  ' Dog mice ! '
they yell, as the large creatures bolt from the track.
Down come the sticks, and the boys gather their
short-tailed, large-headed quarry as they run.  What
with the clacking rattle of the mower, and the click of
the boys' sticks, the mice have a very uncomfortable
time of it.  If we waited now long enough, those
boys would be sure to come with a dozen of the finest,
and ask if I ' wanted 'em for owls.'  I have never
found myself firm enough to refuse these, for the in-

formation had spread round for miles that 'a big man in a grey suit o' clothes, as always carried a big ash stick, one as looked as if he'd bin a soger, he'd actoo-ally gi'n as much as a penny apiece fur mice.' But 'big uns,' as the boys said, they certainly were ; two of them made an ample meal for my brown owl, Friar. Often was I met on my way home from work by some little country toddler. Not a word would be spoken, the child would just hold a mouse up by its tail, receive the penny, and then cut away as hard as its little legs would carry it. Even when my owls were gone the pleasant barter was maintained as long as I walked and worked in this district ; and I always look back on that period of my life with pleasant memories, in which children, birds, and flowers all mingle together.

But to return to my morning's walk. As we enter the woods the sun is high. Firs show their tops through the mist, which now is clearing rapidly away. A magpie appears for one instant, and vanishes ; another flits out, but there is no noise ; nor do we make any, standing as we are on dead leaves which, being wet with the dew, do not rustle ; in certain parts the woods are silent, and now the vapours have cleared off over the tree-tops, and have rolled away.

From a grey thorn a blackbird flutes, as only a blackbird can ; his mate is sitting below him on her nest. This is bird music in perfection, for you have the singer in sight. There he is in full light of the morning sun, his jet-black plumage glistening, and his orange bill showing like a point of light. Stay awhile and listen as he sings his song ; with it you have the life-giving scent of the woods, the very essence of their living growth.

The bird's song rises, falls, and dies away ; the light wanders here and there, now up, now down, on the boles of the moss-spangled giant beeches ; the young golden-green foliage quivers in the light, the branches wave and softly rustle, and the bird's glorious song breaks out again and again.

The author of a popular lecture on ' Music and Morals' would have us believe that there is no intrinsic beauty in the song of any bird ; he pretends that its charm is only due to the force of association. When we hear a bird sing, straightway, he says, we picture to ourselves the bird's surroundings, and so are delighted with sounds which in themselves are not in the least musical. Nothing of the kind. In the matter of birds, I am convinced Mr. Haweis— broad as his sympathies may be in other directions— is a Philistine.

By the way, never look at preserved specimens when you can study the live bird. I have seen owls preserved so as to show three toes in front and one behind, and, though this is quite wrong, some works which the credulous public have been gulled into considering as standard works have also represented them in the same way : that is, in an impossible position. No owl that ever I possessed —and I have kept a goodly number—ever placed three toes in front and one behind, although they could do this for a moment if they chose. Another blunder I must protest against: no owl seizes his prey or holds it with both feet, though both feet may be used to carry it when the prey is a large one ; such quarry, for instance, as a full-grown rat, or at times a pinwire dotter, called by courtesy a rabbit. With one foot the owl grasps his prey, the other foot grasps a tuft or some other inequality of the ground. Then the bird goes to work.

A few owls remain in the wood through which I am leading my readers ; some of the long-eared species, but not many, because when this particular piece of land was let for shooting, a few years back, an order to destroy all creatures that were not game was given. This applied to the utmost limits of the shooting, so that the head of game should count out better than that from the covers adjoining. But what with one thing and another, the man who rented it did not have anything like the head that his neighbours had, much to his own surprise. Careless and indiscriminate trapping and shooting does not often produce good results. Where so-called vermin are popped at all the day long, and the traps set for them get filled with the game itself, the creatures will seek sanctuaries where so short-sighted a policy is not in force.

Wood-pigeons shoot up with a flap-flap-flap; they soar above, where they float with outspread wings and tail like giant tree-pipits; then they settle again; their mates are near, sitting on their nests, or rather their twig platforms. That blackbird must have given the signal for harmony, for on all sides the thrushes sing. Loud above the other voices sound the notes of the missel-thrush. But he sings best in stormy weather; on a calm, golden-tinted morning such as this his voice has too trumpet-like a tone.

And now there is a lull, during which we hear close to us a plaintive hurried song. There is something that arrests your attention at once, giving one the idea that the bird was compelled through some inner stress to sing, yet against his will. We can see him hovering, like a large moth, over the top twigs of a large beech. It is the delicate tree warbler, near relative to the willow warblers and the chiff-chaffs ; only he is of far more secluded habits than they. You may see fifty of the others before you meet with the little tree warbler. He frequents high beeches and great oaks as a rule, where the woods are broken by open glades ; in which spots the delicate wood wren and the tree warbler are not uncommon. The parts which he resorts to are the very ones about which to look for that rare butterfly, the purple emperor.

By taking one more path, through this copse, we shall reach the foot of the fir-covered hill that overshadows the Valley Moor. The new shoots of golden-green and purple are finely contrasted, in this bright light, with the old foliage and the red stems which rise from what might be a thick lawn of whortleberry bushes. These are in flower now : when the fruit is ripe the place will be lively for a time with children and vipers. And without harm to either, for the children know that the reptiles will get away as fast as they can, if they only frighten them a little. You will meet little toddlers, under the super-vision of larger children, stained from head to feet with the deep purple juice of the ' hurts.'

E

We gain the top and look down the valley. And
what a sight is there! One worth going many more
miles to see. A gentle breeze has arisen, just strong
enough to bend and wave the hardly yet matured
masses of green foliage, which look like a sea of
rippling verdure. The fluttering whispers among the
young leaves are distinctly audible.

From the firs comes that aromatic scent of resinous
compound. This mingles with that of the beeches
and the oaks; then there is the odour of the whortle
bushes and the ferns superadded; each scent differing,
yet here delightfully blended, making the finest
medicine a man can take through his lungs.

Before, and below us, is a long stretch of valley
moorland, fringed on either side by woods. Our
readers will please bear in mind the fact that there
are two kinds of moorland, the upland moors and the
valley moors. The former are far more boggy than
the latter, for this reason—the water continually
coming down from their springs forms the beautiful
little trout streams that drain the valley moors.
From rills, from mere threads, and from drippings,
falling drop by drop from the mosses, the water
comes now as it has done from time beyond record,
from the moors above to the moors below; in summer
and winter a continual flow, without stint. Now, as

we stand on the crest of the hill the water is trickling down below our feet through the moss. At the foot of the hill, separated from the wood by a strip of bog, covered by great clumps of rushes and cotton grass, we are expecting to see some bird life, if we are very cautious ; but we must avoid that bog, it quakes. I have been in it once, only at the edge, but that was quite enough. A grey alder stem, grey with moss, sticks up in the middle. That is the only object that strikes our eyes yet. I never do expect to see much at one time, and even if we see nothing we should not fret ; we shall meet with all we want at some time or other, if we live long enough. Sometimes, for weeks and months together, I have not seen enough to talk to a child about, if I had wished to amuse one ; and then again have suddenly seen, at a time when I was not looking for it, more than I could have hoped for.

If any particular spot has been frequented by any creature, furred, finned, or feathered, and the spot remains as it has always been, unless the species has been quite exterminated, we may look for the same creature there when the season recurs again. I have put blackcock up here many years ago, one of my woodland friends having invited me over to see them play up. They are very partial to rushy flats, and feed in such spots more than on higher ground.

Just as we are about to descend the hill there is a commotion in the rush clumps, and a bird tumbles out which we at first sight take for a wild duck with its wing broken ; but the cry quickly sets us right. It is a grey hen—a mother—for we can see some of her poults dash to cover, the mother with them. After that all is still ; not a sound or a cheep to be heard.

Looking over the bog, something on that dead alder catches our eyes ; it is a hen sparrowhawk, sitting as motionless as the tree itself. Sitting there, as upright as a drill sergeant, she might easily be mistaken for a part of the grey trunk. She has missed her quarry ; so much the better. The first movement on our part, as we wish to descend, is enough ; and with a flirt of her broad tail she is up and away. Not for long will she have the pleasure of providing for her young hopefuls, now of some size, for her domicile is known to our friend. She has had some of his young guinea-fowls, cute as they were ; and for that he told us he would knock her out of it, and the young ones too.

From the bog the water has cut a natural course for itself through the wood ; and, as it is all down hill, the water runs from a clear pool in the centre with some rapidity. It would have formed a natural cascade on entering the moor from the wood, if some one who farmed the lands near had not provided a very ingenious outlet for it. This was formed by the half of one of the trees which at one time had stood there; it was hollowed out to make a spout for the water, and rough scantling of great thickness had been placed on the top; then turf, more scantling, and then turf again over all. The run of water was then directed into it, and it runs there still, though the man who effected it first is forgotten. Like other jobs of a bygone day, when there was less hurry, it was done well, and that primitive conduit still fulfils its purpose.

An R.A., now gone over to the majority, painted this spot with loving care. He called it 'The Moorland Spring,' a most fitting title. Lush herbage and tangle grow all round, almost hiding the spout from view. But you see the great flow of pure crystal water falling into the pool of the little trout stream below. Here the chattering, scolding white-throats, the greater and the lesser, gather; something about the spot suits them, and also the nightingale; in the season you hear his melodious singing close to the spring, both by day and night.

Pigeons flew down to drink also, and the black-birds, thrushes, and finches bathed there. It was, and is, a beautiful bit for the artist. Into the picture to which I have alluded a cottage girl was introduced, with one of those old-fashioned brown pitchers in her hand. She is about to fill it under the spout; and this gives just the human element that was needed.

The stream runs swiftly and merrily on, forming picturesque miniature bays. Here is one where an old moorland stump has tried to shoot out again and failed. Though tiny shallows, some of the bays are six feet in length; then comes a pool the size of a small table top; after that a run of water two feet deep for twenty yards or more, and so on, a complete change at every turn; at the widest part it is not six feet in width, even where the cattle come to drink.

On it goes, making rippling, lapping music, all through the moor. Here the titlark, the meadow pipit, nests in the tussocks that rise up from the spongy parts. The bird sits close; if you are careful you can see all you wish without disturbing the inno-cent creature that is watching every movement of yours with her bright eyes.

Pewits breed here; three of them are now, to all appearance, unconcernedly feeding quite close to us. They are, however, simply watching every step we

take as we jump over the stream on to their part of
the moor. One is suddenly taken very bad indeed,
as we say ; something appears to have got him by the
throat ; he tumbles and flaps so much. No need for
all this, we shall not go nearer to them. They are
satisfied as to this, and springing up, the signal is
given, a sound which is at once answered from a field
close to the woods. Over the splashed moor bank
they come, flapping, wheeling, diving, tumbling, and
shrieking Pewit-wit-wit, Pewit ! Their young are
about, or their eggs are in the field ; the first three
were only pewit scouts. The birds are not molested
here ; even the rooks have had notice to quit, so that
the pewit's eggs may be let alone. The rook's food—
his legitimate food, at least—is wire-worms, grubs,
worms, and beetles, and for poaching on the preserves
of others he is punished here ; we notice a couple of
his family spread-eagled out on the ground. If wild
creatures are allowed to indulge in luxuries, the liking
for them soon passes into a necessity, and becomes a
rooted and transmitted habit. The kea, or, as it is
generally called, the kaka parrot, thus acquired its
taste for the kidneys of the live sheep—a perfectly unnatural one.

Meadow pipits cheep and run about not very far
from where their mates are sitting so closely, hidden

in the bottom of the tussocks.   And the air is all
alive with tree pipits, or rather with their voices; they
rise from the lower and outermost boughs that reach
out over the edge of the moor, right over the very tops
of the trees ; then the wings and tail are spread out,
and the gay-hearted little birds gradually float down
singing, to the very twigs they started from.   This
is repeated over and over again, a rising, floating,
musical performance.

Great green dragon-flies dart past with their wind-
mill-like sweep of wings.   You hear the rustle and the
click of them as they turn in flight ; wood-pigeons
shoot over from the woods to the fields, and back
again, whilst a plaintive Coo-coo-roo-roo-coo Cooee
reaches one's ears, and sounds wonderfully conducive
to repose ; to which, however, we may not yield ; nor
do we wish to, we want to see all we can, as we do
not come here too often.   Beside a little shallow, just
below, we come on one of the most beautiful sights
that a wandering naturalist can see—that of a pair of
those graceful little water sprites, the grey wagtails, in
their own home.   The colours of this lovely species
are as pure and bright as the water and the sand that
he wades in and runs over.   Some of my readers
may not have seen the grey wagtail in his full breeding plumage.   The head and back show warm grey,

with a tinge of green ; there is a white stripe over the eye and one below ; the throat is velvet black ; the upper tail coverts yellow ; the breast and under tail coverts a rich golden yellow ; wings and tail warm brown. A few of the tail feathers are edged with buff. Form, colouring, and movements are alike beautiful. As he sits on one of the stones of the stream, his head up, his long tail just touching the water that is rippling round, the bird and his surroundings would be sufficient for a good-sized canvas. His mate is daintily tripping on the edge of the stream ; happy, blithe little creatures are they, in this moorland home.

Often, when I am wandering alone, some who have been my companions in years gone by come to mind, unbidden guests of memory, or called up out of the past by some trivial incident of the present. ' Father Jemmy ' was one of these. He received this nickname before he had been one week at school. His father brought him when they came to settle on their farm. Never shall I forget the pair as they entered the schoolroom together—forty-three years ago it was. I had long ere this done with school, but I happened to have some business in the building that day. A sturdy little figure was Jemmy, a man in miniature, wearing good stout shoes and

leather gaiters of a dark brown ; these reached to
his hips, having no end of buttons to them. A short
round frock covered his upper person. The father
was dressed in precisely the same fashion, they seemed
a large and small edition ; even the shirt collars were
the same. I knew Jemmy, too, when he had done
schooling, and had grave matters to attend to ; for his
father died, leaving the mother to his care. Mother
and son understood each other perfectly, and he was
devoted to her. The measured words of premature
wisdom that fell from the lad's lips gained for him
the title of 'Father Jemmy.' But what a shot he
was ! He could fish, too, with a rod and line, or
without one ; and, rarest gift of all, he could keep
his mouth shut when there was not any good reason
for opening it.

Quaint, manly Jemmy, he did his best for his
mother as long as she lived, and when she died he
was not long in following her. A fine eight-day
cuckoo clock stood in one corner of their kitchen, inside
the case of which he kept his single-barrelled gun ; a
good one it was, and always kept ready loaded. 'I like
to have things ready and handy-like,' he was wont to
remark ; 'you never know when they may be near'—
the birds he meant. There was no gun licence needed
in those days, only a game certificate. The things that

did come when Jemmy was about seldom went away
again. The clock was a good one, but the bird had
not shouted properly for some time ; one day he
would tune up all right, and the next he was mute.
Mother said she really must have it cleaned, but she
never could make up her mind to let it go out of the
house. One day, the very day when this really was
going to be done, a boy came rushing from the
farmyard into the kitchen, where Jemmy was having
his snack of lunch alone, to tell him 'Something had
come, and he'd best be quick, for it was on the move-
like.' Jemmy opened the clock-case, and hurriedly
caught up the gun, as he had done scores of times
before, but not with the same result. In the haste
of the moment he touched one of the weights, it
swayed, the hammer of his gun hit the bottom of the
weight, which threw it back, and off she went, Bang !
right through the works. Down fell the cuckoo.
The mother was upstairs when she heard the report ;
her first thoughts were for Jemmy. 'Jem, my boy,
what have you done ?' she cried. For a moment he
looked speechless at the wreck, then he shouted
back, 'All right, mother, I've only cleaned the
clock !'

To return to our creatures : having permission to
fish I may as well avail myself of it, for the stream is

alive with trout. They give no sign, however ; the water runs too sharp and rippling to show a rise, and the formation of the stream would not allow of a fly being used. So much the better ; my preparations for taking a brace—merely to study their colour, for I do not care for fish diet—are soon made. I take the first three joints off my walking-stick rod ; then I fix up a fine gut line, one foot shorter than the three joints ; no shot on it, only the hook. From my very small worm-bag I take a nice red worm of only medium size, and place it on the hook. Now I am ready ; here we fish down stream. There is a large stone about six feet below me, round which the water spins. Gently stepping back, I throw about two feet above it ; the worm goes over. Tug-tug, snick ! I have him, and he is on the grass ; a small fish, but very thick for his length, which is about that of a small herring, and having a golden back and sides crimson spotted belly pure white.

On a shallow below, where the water barely covers the stones, more for the fun of the thing than anything else, we pitch our worm. The water has barely closed over it, when something shoots out from the bank and back again. Snick ! comes out of it another trout ; but he is not so golden as the first. He is not hurt, so back he may go again. If a

fish has been handled, kill it ; it is only right to do·
so, in order to prevent its future suffering ; for scales
rubbed off bring misery to fish.  When only just lip-
held, it is another matter.  My next throw is above a
small hole where the water spins round.  Hardly a
moment elapses before there is a tug, and we strike.
This is a larger fish, and he fights well.  Half a pound
will be all his weight, but he is a nice fish for this
particular part of the stream.  It is a somewhat
curious fact, but if this stream flowed into a lake or
pond, the trout would in the larger water become
large fish ; and if large trout found their way up a
stream of this kind from a pond or lake, they would
very soon dwindle down to the size of the pair I have
just captured.

Having caught my brace, I make a sketch of them ;
and the first boy or girl I meet on the way home will
become the happy possessor of the fish.  Generally,
when I want to paint fish I have them brought to me
alive.  By giving sufficient notice I manage to get
these just when I am ready for them.

As I near my friend's farmyard, I pause for a few
minutes to look into the mouth of a large culvert that
carries the stream under the main road before it runs
through the farmyard.  The water at the mouth of
the culvert is about three feet in depth.  A small bay

has formed here, and the water is certainly suggestive of fish.   I had put up my tackle, but I must see what there is in the culvert under the road.   So from my bag I take five or six worms, and pitching one right in the mouth of it, I wait.   No sign ; I pitch again— all quiet.   The third I pitch a little on one side of the mouth of the culvert.   There is a rush, and a boil up, and the worm is gone.   After this we drop them in the main current, about a yard from the mouth.   Out shoot a dozen fish, with a rush ; first-rate fish.   Having learned all we wanted to know, we can pass on.

Wild ducks, both the pure wild birds and the half-wild ones, breed up in these moors.   It is impossible to tell which is which.   They nest on knolls and heathy kobs, the very driest spots they can find in the heather.   Water is all about them, but their nesting places are dry.   After many years of patient watching, I have come to the conclusion that this is to preserve their young ones from the numerous enemies that are on the watch for them at that time, until they are quite capable of taking care of themselves.   They know when the time of danger comes round, and they watch for the foe above and below.   Feeling they are safest away from the water, they nest in the haunts of the blackcock and the ring-ouzel.   Who that did not know the habits of the creatures would ever expect to find the wild ducks' nest in fir and heather? or to find a nest on the top of a large sand rock, cosily placed in the hollow of a beech stub, close to the home, or earth, of a fox? Anywhere, it would seem, except near their own quarters. For murderous rats are there. No matter where it may be—in the covers or on the lakes or ponds, even on the bleak foreshores, wave-washed and wind-swept—you will find that pest. Wherever man takes wild creatures under his care— or thinks that he is doing this—the rat follows him, to see what kind of a job he makes of it.

If wild fowl that are visiting waters get fed, in order to encourage them to remain, the rats will come to see how they eat it, and contrive among themselves to appropriate the food. No job is too hard for a rat, and I can say from personal experience that a past master or mistress in ratcraft it is hard to circumvent. The common house-rat, the brown one, is a first-rate swimmer and a good diver. He will watch the ducks and listen to all their calls, and if the birds have located themselves on some small island he swims out to it, steals the eggs, or kills and eats the young birds. If the nest is near the water, by the side of some stream or outlet, the rats will cross with the young ducks, after they have killed them, to the other side A dozen young ducks, half-

eaten, have been found under one large stone close to
the water's edge. They will watch, close by, for the
departure of man or dog—they know well when
these have gone—and then they do their work quickly.
Stoats and weasels are really benefactors to man,
for they will kill rats and mice in great numbers.
Pheasants and partridges suffer in like manner as the
ducks, but to a less extent. A lot of wild ducklings
with their mother is one of the prettiest sights
possible ; the mother's watchfulness, having to look
out above and below, is very interesting. You will
not see the pike close in shore, under the weeds, but
it is lying there, near where the little birds are
paddling. I once let a half-fledged sparrow fly out
of my hand, by accident, over the water. It settled
on the weeds and fluttered a little ; the weeds moved
from below, there was a slight opening in the clear
water, then a plunging splash, and the sparrow was
gone. Pike that have been shot when the ducks
paddled have called forth some stiff language from
the farmer, when their insides were turned out.

I have seen mice on the moors, but I have never
seen a rat there. The foxes would have them if they
tried to settle there. They are found in the fields in
numbers during harvest time—fields that are at the
foot of the moorland hills ; but that is as far as they

come. In some way the knowledge is transmitted that the wild land of which man can make no use is the safest place for the creatures to nest in ; the birds seem well aware of this—the wild ones know it, and the half-wild ones find it out. Indeed, man himself is at times best off out in the wilds, apart from his fellows.

Under the tender green leaves is the place to watch the birds nesting, or rearing their families. For more than half a century I have had this pleasure each year, and I hope to have it for many a year yet to come. What care and self-denial is theirs, as they feed their young before taking their own food! Ever ready, too, they are, to sacrifice their lives in trying to defend their offspring. And the birds of prey are as gentle in this respect as is the robin. Reynard himself is a good father, and his vixen partner a most exemplary mother. Stoats and weasels will carry their kittens and will defend them to the very last. As to the hated rat, no living creature fights a better battle for its young. Much, indeed, there is to instruct man, and many matters for him to ponder over out in the open, when leaves are green.

F

IN an article entitled ' A Roadside Naturalist ' I have
tried to show that to the country roads field naturalists
are indebted for a very great amount of their insight
into the life of wild creatures and their ways.   From
our main roads others branch off, leading into the
Weald lands of Surrey and Sussex.   You can go a long
day's journey, rest for the night, and take a still longer
walk the next day, all through green lanes.   As they
have been for many generations, they continue to be.
In one respect they are different.   The roads now
are fit for traffic—this is the only alteration ; for the
land is as it has been for centuries, and the same old
manor-houses and moated farms are still there.   Forty
years ago these green lanes or roads at certain seasons
were impassable ; even now the posts which recorded
the water's depth remain by the side of the bridges
that span the woodland trout streams, which, in wet
weather, change from trickling streams to rushing
torrents.   So little was known about this network of

roads, green though they were, that, at the time we have mentioned, the inhabitants of country towns within a few miles of the edges, so to speak, of two counties, knew nothing about them, beyond the report that such places existed. What little was heard at intervals, which were few and far between, was always connected with smuggling. For months, at different periods, I have made my home in various parts of this quiet, lonely district, full of beauty, go where you will.

It was an open secret that smugglers crossed the Weald from the south coast, and left portions of their runs in various parts of it. Smuggling and natural history are very different subjects, but it was through being shown the places most used and frequented by those who followed the former pursuit in past years, in very secluded spots, by men who had known something about it, that I first became acquainted with the wild things and their haunts in the Weald lands of Surrey and Sussex. Those who are sons of the coast have very little difficulty in making friends with others who have known coast people ; and so pleased was I with the kindly and hospitable people of this wild district and their homes, that from time to time I returned to wander there. When I first knew these districts no railways existed. I believe that they had not even been thought about. Indeed, as matters are at present, you have to walk four or five miles before you reach a railway station when you are in the centre of the Weald. Eighteen years had passed since my last visit, yet, with the exception of a few new mansions—not many—and one new shop with plate-glass windows, in what can only be called a village by courtesy, things remain as they have been for centuries, so far at least as outward appearances go. If our readers will recall to mind the sudden formation of railways through districts that had never had anything except the plough to move the earth—all this within the last twenty and thirty years—they will be able to form some opinion of the wonderment of some of the old folks in this out-of-the-way region when they first heard of the railroads being made in their midst. But their existence has made little change here. The people still have many primitive customs. Take, for instance, one of these before the change came. The farming portion of the population—nearly all were of that class—paid in kind ; very little money passed. If the village shoemaker made shoes for a small farmer and his family, he generally took it out in farm produce. This system still prevails more or less. I have read much that has been written by those who have pretended to understand the agricultural population

of the southern counties. Their writings seem to me
to prove very clearly that they know nothing whatever about them. Slow of speech Hodge of the Weald
may be—that is a virtue, as a rule ; but his wits are
keen, and those who, for want of better knowledge,
think they can easily enlighten the 'tiller of the soil,'
as they term him, will find that he can enlighten his
would-be instructors in a very unpleasant manner. I
am personally of the opinion that the class to which
he belongs forms the very backbone of England.
His reticence is not dulness—far from it. Before he
takes you into his confidence he wants to know who
you are. If he thinks you will do, he will show you
anything he thinks will please you, as soon as he has
found out what your hobby is. If you begin by asking a favour of him, he will, in most cases, flatly
refuse it.

The lives of these people are passed in the open
air. Both summer and winter they are in the fields
and about the woodlands that stretch for a great distance in the forest districts. This country is well
wooded right down to the sea-line. There are great
farmsteads with moats round them, the roofs covered
with slabs of stone in place of tiles or slates, the path-
ways leading from the great covered porches right
through the orchards and, in many instances, out to the

green lane or road. These paths are raised, and paved with the same kind of stone that the roofs are covered with. Old-time houses they are, divided and subdivided by great oak timbers, which are fixed into the brick and stone work. Silver-grey the oak is, but solid and firm now as when the timbers were first placed there. Huge chimney-stacks there are, with a large slab of stone on the top of the chimney. Four piers are built up to carry the slab, so that square openings are all round the tops, giving them a very quaint appearance. Wood only was burnt in former times, and in some of these old farms wood only is used still. I have seen great logs placed on the iron dogs. When these were well alight, the stone-paved kitchen or living-room, with its great beams crossing the ceiling, glowed again.

This is a land of birds and flowers. The whole of the district in the months of April and May, the roadside stripes, mile after mile, the hedgerows, copses, and the woods, are gay with bloom. In July the modest wild flowers—such as primroses, blue-bells, the wild hyacinths and anemones, and many more— have given place to the gay flowers of the wild lush-tangle, all that grows in moist or dry places. The vegetation of dry lands and of the swamps flourishes here in the rankest luxuriance. Step off the made road on to the green stripe, and you at once sink in wet ground ankle-deep. The fields that rise above the road, in most cases, either on one side or the other, are dry, very dry, for it is a clay soil here. The tangle is as luxuriant as that of the swamps. Tussock or hummock grass, great clumps of it, and sheaves of tall rushes, grow close to the edge of these roads, which were made over what was simply at one time, within my knowledge, an impassable swamp in winter. Thousands of faggots have helped to form the foundation of some of them. If you look at the cattle that at times graze on the green stripes, attended by one of the farm lads, you will see mud to above their knees. Their weight breaks through the thin dry crust on top. If any portion of the cultivated lands here were neglected, they would in a year or two go back to wild land again. A sturdy race are those dwellers on the Weald, and they need be. Puny folk could not do their work. Why certain political agitators should, for their own purposes, attempt to make this stalwart class of men, with their sturdy independence, and loyalty to their employers—we will not say *masters*—pose as a 'down-trodden' or oppressed class, is only known to themselves.

The hearts of the farming class are in the right place ; and we know that all their sympathies are

*F 4

really with their old employers and with past times. All things change, but the change from the old *régime* to the new is, in many instances, a violent one. When, from some cause or other, large estates have changed owners, things have not worked well. It could not be expected that they should. The farmer and his men fully appreciate and return any act of courtesy. They are gentlemen by nature. I have in my lifetime had opportunities of studying a considerable amount of so-called society, some of it as high as was to be got, and some of it ' low '—so called because its members had not so much money, and I have known as much good-breeding in a woodman and his wife as could be seen elsewhere. Talk of field naturalists ! Such men are this in spite of themselves. Works on natural history, provided especially for the use of village libraries, are things to mock at. A lie, no matter how it is dressed up—and some lies are very nicely dressed—*is* a lie. Men who have the living creatures before their eyes, their means of living, and their various ways of procuring it, require no books, not even the scientific ones with their errors. Fifty years in the open air, in all weathers, will teach one much. Some of the men I know have passed eighty years, and they are yet well and hearty.

Their knowledge of all things that shelter beneath
green leaves is very great ; in fact, it is the knowledge of a long and necessarily observant life.
Nearly all the cottages have orchards—the fruit-trees
in some instances trail over the hedge that separates
the garden from the road. Great boughs, loaded
with fine fruit in the season, lie on the hedge,
mingling with the fine blackberries. If you see the
owner or his 'missus,' and ask if you may have an
apple or pear—to offer to pay would be considered a
direct affront—permission will be gladly given. He
or his wife, as the case may be, will get you two or
three from the top boughs, where the sun shines right
on them.

A collector of insects cannot expect to be received
very favourably when he goes about round the
borders of covers at night with a dark lantern,
sugaring trees with treacle and strong ale to capture
intoxicated moths. One I knew had to run for
dear life ; for the keepers, seeing the flashes of light,
now here, now there, for the first time in this district,
thought it was some new poaching device that
was being practised. When the entomologist was
walking to a fresh lot of trees on the edge of the
next small cover from the one he had just finished
exploring, he heard voices stating, in no measured

terms, what they would do to the 'poachin' warmint'
that 'was at it with lights.' Very gingerly he crept
over the soft green stripe in the opposite direction to
the sound of their voices. Though most innocent of
bad intent, he knew it would be a difficult matter to
convince those who were looking at that light that he
was so. He remembered too, like a flash, that he had
been told at the house where he put up that no one
was allowed in this particular place in the daytime,
and less still at night. The truth was, a particular
moth that he wanted could only be found on some
tangle which grew on the ditch-banks that surrounded the covers. He fled, and all went well for
about one hundred and fifty yards. He was on the
high road, about one mile from his lodgings, when,
owing to some hurried movement he made, he accidentally for a moment turned the light of his
lantern on in front of him, up the road on which he
was running. It was more than enough. He heard
a voice roar out, ' Damn 'em ! there they be ! ' Then
came the bay of a great hound in leash. The underwood crashed, and he fled at top speed. His
specimen-boxes flew out of his pocket, full of small
precious moths. In tearing at some of his buttons,
so that the air could play on his chest as he ran, the
lantern was pulled off his belt and dropped. This

the hound nosed when he reached it, a fortunate circumstance for the man—although he did not consider
it so at the time—for it caused a slight delay.
Dashing round a corner, he cleared a low hedge, ran
across a meadow, and gained the back part of the
small inn where he had put up, breathless. After
cooling down a little and making himself presentable,
he had a glass or two of ale, and went to bed. The
next morning, as he sat in the neat little bar-parlour,
turning over in his own mind whether it would not
be better to clear out and visit new pastures, one of
those nondescript individuals that are to be found
on all estates where game is preserved strolled in with
a couple of beagles. To all appearance he was
merely exercising the dogs, but in reality he was in
search of information. The man was dressed in
keeper-fashion, although in the strict sense of the
word he was not a keeper, for he had not reached
the gun-carrying stage.

After some commonplace remarks about the
weather, the landlord, who knew him, asked if he
would not come in and rest a bit, and wash the dry-
ness down. With a half-muttered protest to the
effect that it ' wud hardly do if the " head un "
ketched him there,' he walked in, and after he had
seen the bottom of one pint-jug and ordered another,

he remarked to the landlord, 'Ye ain't seen no strangers, hev ye?'

'No, only this gentlemen here; and you can't call him a stranger hardly, for he's been here nearly a week now.'

'No, I don't mean the likes o' him. I aint a born fool, not quite, I hopes. Fill this here pint agin, my gullet's like a lime-burner's, all through last night's work. 'Twas a rum go that! Our " head un " goin' roun' the little covers—you knows 'em—sees some smirches o' sticky stuff on some o' they outside trees. Well, he don't say much, but he knowed summut wus goin' forrard as hadn't ought ter be, so he comes and gives orders fur to watch them covers. We wus at it last night.'

'Did you get anyone?'

'No, but there was a lot on 'em with lights; artful as sin that lot was! We could see their lights, and I think we should 'a had one on 'em, if the bloodhound hadn't nosed about at one o' their lanterns what they dropped. Old Jack had him in a leash fixed to his leather belt. Squire says he ain't tu let him run loose on no 'count, an' when there's a tussle he ain't to be loosed unless things gets real desprit-like. Old Jack feeds him, and looks after him; an' he's the only one as ken handle him. Last night,

when he nosed that 'ere lantern, Jack said we must
stop, fur if he tried to get him away afore he'd considered matters over, he'd get wild-like, an' there'd be
a job.  It waunt long he nosed it, but it was jest long
enuf fur the one we was almost on to to git clear off.
You see, there was a good many o' them 'ere lights,
an' they must 'a bin about the artfullest lot that ever
come about these 'ere parts, for we only seen one o'
they chaps.  Harry, he picked up a gimcrack thing
o' a box with little owlets in it, with pins stuck through
'em.  You may depend on it, that 'ere stuff on the
trees and them 'ere owlets on pins was to draw the
birds down arter they'd roosted, so as they culd pick
'em off with their hands.  We be goin' there agin to-
night, an' if they comes, we're sure to hev 'em, fur
there'll be more on us at it.'

Exit man, to the edification of the entomologist.

I have known one instance where a zealous collector 'supped sorrow' for having placed a white
sheet upon poles at a cross-road on the edge of the
woods.  Opposite the sheet, on stakes, he had fixed
two very bright lanterns, with large reflectors.  His
movements, as he dashed about here and there with
his net, to capture the insects that were naturally
attracted to the lit-up sheet, caused some strange
figures to be thrown on it, in what appeared truly

magic style. Some travellers that way turned back, and got home by another road ; and one individual I knew, carried for some time a large horse-pistol with a flint-lock. He told us 'she had got a double charge in her.' It was a mercy 'she' never went off near that entomologist, for the man had been so frightened by the wild figures thrown on the sheet, that several times in my hearing he spoke rashly, and declared he would shoot the first thing that came in front of him in the night-time. When the real truth became known, the entomologist had to clear out of the neighbourhood.

A blue sky without a cloud is overhead. All distant objects are shimmering in the heat, which appears like a soft grey veil over everything. The fields and trees, and the farms in the distance, look faint and dim. In front is a long lane without a bend for more than a mile, covered in for the greater part of the distance by the hedgerow growths. Not one house is to be seen the whole length of it. We have already passed through several very like this, but with more houses showing. The bits of landscape we see through the openings in the hedges, where the gates are leading into the large fields, would be very beautiful peeps to place on canvas. It is very quiet, for the birds have done singing. Their young ones

are out of the nest, and they have to look after them
now with the greatest care ; young birds have so
many foes that watch for them and kill them if the
chance offers.   Though they do not sing, they are by
no means silent, for you can hear the calls of old
birds and the answers of the young.   If several different species are about, this kind of conversation is
very animated.   When there is a lull in the birds' conversation, you can hear the whispering rustle of the
wheat—you may not see it, but you can hear it—
whisper, whisper, rustle, rustle, rustle, as you pass
along.   Now and then a lark soars up for a short
time, and drops into the wheat again, without attempt-
ing to sing.   His second family of 'game-looking'
young ones are not far away from the spot where he
dropped down.   The sweet scent of hay comes and
goes, as some slight current of air wafts it to and fro,
from meadows you are not able to see for the wild,
high lush-tangle of the stripes and the hedges.   Huge
docks, burdocks, teazles, or, as they are called here,
kixes or kexes, hog-weed, and the so-called wild pas-
mut (parsnip), with many other plants, tower up high
above your head.   As to the hawk-weeds, they blaze
out in magnificent clusters of orange and pale gold,
while purple thistle-blooms top and nod over all.   If
you wish to see tangle in perfection, you will find it

here. Great dragon-flies, green, orange, and blue, are abundant. They rush to and fro, now high up, now low down. These damp stripes are their hunting-grounds, for their mixed prey is legion here. The hum from their unseen wings sounds all around you, above and below. The swallows that nestle about the farm and the farm-buildings twitter as they dash about where the hum of insect life sounds loudest. That dreamy, sleepy hum from wings unseen, at times penetrates the ear like a faint music of subtle har-monies. Shrikes revel here, and scold in the most violent manner, if they think you are getting near their young ones. Jays have a decided taste for hedge-life about this time, but missel-thrushes object to their presence. Mice the jay kills and eats, but the thrushes know that if they were out of the way, he would have one of their speckled-breasted young whom they are trying to teach how to get their living. Occasionally you hear a low croak—this is the warning of the nightingale for the young ones to keep close cover. Butterflies flit in all directions, and the willow-wrens and white-throats slip in and out as you pass along, from the sallows that in places line the edges of the green stripes.

You will not go very far before you pass over a bridge, and that will be, as a rule, close to a farm—two miles away, it may be, but that is not considered any distance in this district.

'How far am I from the next village?' I once asked.

'You are close on it,' was the answer; ''tis about two miles, or a little over.'

From the top of the bridge to the bed of the trickling stream below will be from ten to fifteen feet. It is the natural channel that the water has cut for itself through layers of clay and slabstone. The marks on the water-posts reach up nearly to the belly of a cart-horse. This is the safe-distance mark. A white bar projects for all to see, and when this is covered no one attempts to put a horse through the stream.

It is very pleasant to look over one of these small bridges of one arch only, in midsummer, and to watch the trout shoot over a shallow, with hardly enough water to cover his back-fins, into some hole beyond it. But it is a different matter when summer has gone; for I have seen the bridges covered, and only the tops of the white posts showing above water, and a rushing torrent, both wide and deep, passing through the woodlands, making the tops of the young trees bend and sway with its force, and sending the kingfishers, which are numerous here, and other shy things, close to the roadside. I have seen kingfishers perched

G

in strange places ; for although the bird gets his living
from the water, if he is not able to get the rest he
needs, he will drown. It is possible even to drown a
duck ; all birds must rest at times.

The fields and the water-meadows are very large,
acres on acres of hay, wheat, or root crops ; and when
these are not on the land, and the fields do not touch
on covers, permission is freely given to prospect about
in the study of wild life. What a grand sight it is to
see a mass of thunder-clouds top the ridges of the
South Downs, fifteen miles away, and sail over the
edge of the Weald ! There is no blue haze now in
the distance—that has been carried off by the current
of air that forces the clouds along. These gather
until they appear to rest in the very centre of the
Weald, over which we are walking. A glorious sight !
Huge clouds, mountains of them, pile upon pile, seem
to have their bases in the fields, and their crests above
you, ready to topple over. There is the great expanse
of the flat all around, with the hills that surround it in
the distance. The cloud-masses are a dark purple-
grey, with lurid tops of light buff. Here and there
you will see great ragged streaks of steely grey, that
show for a few moments and vanish again. This
weird light throws all things up in strange relief.
The distance is seen in detail ; as far as the eye can

reach, you can see the general outline of things. The bolder part of the clouds gets darker in colour, almost inky, and the light edges are lit up with a brassy light. It looks as if one were standing under a vast dark dome, with the light coming down through a rent in the top of it. As the air moves the corn, which is just beginning to ripen, the golden-green ears show like points of light as they bend down and spring up again. The brush-willows sway for a moment, and are all in a flutter, their pointed grey-green leaves quivering and spinning. Then all is still, there is not a sound, not a chirp. We walk into the yard of the manor farm to see how this gathering of clouds will affect live creatures. The swallows are all right, for you can see them poke their heads out from their resting-places. One that has its nest on a beam over our heads flits down, and flutters the length of the shed and back again, like some great long-winged moth, so gentle is the bird's flight. This is merely to see if we come with any ill intent. Being convinced in his bird-mind that we are quite harmless, with the feeblest of twitters he sinks down into his nest again ; for swallows use their nests as resting-places when their young are on the wing. Flocks of sparrows rush up with a whirr, and dive into the holes in the thatch, from which refuge we can hear them congratulating themselves

G 2

that they have reached it in time. There is a 'chip, chip, chip!' and a 'chissick, chissick, chissick!' from the whole community, then they are silent. Geese and ducks come from the pond, and squat down underneath the waggons and carts below the shed —their heads turned on to their backs, their bills buried in the feathers. The poultry come filing in silently—not a sound do they make—the hens first and the roosters bringing up the rear, combs lowered, hackles pressed close to the neck, and the tail-feathers—the sickle-feathers usually held up so jauntily—trailing their tips, in some instances touching the ground. One solitary guinea-fowl, that had roamed away from its companions, comes rushing across the yard screaming out its cry of 'Come back, come back, come back!' The dog is curled up in his kennel, and the cows are standing close together under the cow-shed. Now all is silent, saving the 'pit, pit, pat' of large heat-drops of rain. As we look out, a flash of forked lightning shoots right across the hollow of the closed dome. The thunder follows instantly, with a crash that makes things rattle. This continues for some time, then the rain falls in a gentle shower, the greatest boon for the wheat. A breeze springs up, and the cloud-masses break and drift. They are sailing away from us, and we have only had the beginning of it, for we can see the rain descending like sheets of vapour. Leith Hill and the hills that follow, on to Hind Head, stand out in bold relief for a few moments, and then they are blotted out.

The sun shines ; the drops of water on the leaves of the great vine that runs up the side and over the roof-slabs of the house, glitter like diamonds. A pair of starlings fly up to the top of one of the chimneys, and chatter in their quaint fashion, with quivering bills and puffed-out feathers, about the change of weather that came and went so quickly. The sparrows have left their holes in the thatch, and gone back to the spot they rushed from. The geese and ducks waddle to the pond again, where they throw the water over their backs, flap and splash with their wings, and have a general trim-up. The roosters stalk out, give themselves a shake, and there they are in all their bravery—combs and tails up—followed by the hens. They crow now. With a ' Come back ! ' the guinea-fowl darts across the yard to finish his exploration in some hedgerow that the storm had forced him to leave.

Swallows come out one by one ; presently they are dashing about all over the place. The one that was on the beam overhead dives down and out to join his

mates. From the fields, where they have been sheltered
in the thick hedges, linnets, greenfinches, and yellow-
hammers fly up to branches of a dead ash, close to
the farm. They have much to say, for they chatter
and warble in the most pleasant manner ; and as you
listen you feel that life is a very pleasant thing after all.

The farms—some of which, the moated ones, have
at one time been gentlemen's houses—are very old-
fashioned, very solid, and very comfortable. The
great stone-paved kitchens, in past times the servants'
halls, are cool in summer and warm in winter. Well-
conditioned they are, surrounded by woods, fields, and
water ; and the waters are well stocked with fish, not
artificially. Mother Nature is most generous and wise
in all her works, and she does her own work here, as
it has been done from the beginning of all things,
perfectly.

The large cottages—they are called that now—
built on the same plan as the farms, were, we think,
at one time occupied by the headmen on the farms,
the bailiffs of the gentlemen who lived in the large
houses. All the lower rooms of these cottages are
paved with flag-stones. This answered very well for
the generations for whom they were built ; but
these are gone, and their ideas with them. Most of
those stone floors now are covered with a matting of

some kind, if the inhabitants can any way afford it. Women now will not wear hob-nailed shoes, as they did in past times. As one old lady observed to me, ' Flags is all werry well, but a boorded floor is a deal comfortabler.'

I have roamed along those long green lanes in summer-time, from sunset to early morning, listening to the voices of the night. It is fresh then, the moist stripes make the hot and dusty roads feel cool. Sedge-warblers—not reed-wrens—chide and chatter. If you throw a stick into the tangle they will carry on at a rare rate. That swish of wings comes from wild-fowl of some kind, but not wild ducks ; for they are hiding in the water-tangle finishing their moulting, getting rid of their old feathers. They will not be able to fly well before the acorns fall. A clickering chatter is being carried on, like the chatter of an owl when he talks to himself as he flaps along. It comes from moor-hens settling some domestic question in the pools where the rush-sheaves flourish.

That whirr, breaking out at intervals, something like the run of a pike-line off the reel, proceeds from the grasshopper-lark or grasshopper-warbler. ' Crake, crake, crake, crake, crake ! ' comes from the land-rail, or, as it is more frequently called, the corn-crake. This is answered by another. They frequent the hay-fields more than other places. Most of their food is procured there; a certain amount of moisture is necessary for their well-being, and this they find in the meadow-lands.

The droning 'chur' of the heave-jar comes on the ear; but you must go to a fern, fir, and heather district to get its spinning-wheel music to perfection. Oaks take the lead as woodland trees here. As a rule, birds do not sing very often in the heat of summer. As a rule, I say, for want of a better form of expression; for really in bird-life there are no rules: they are affected by matters that do not concern other creatures much. Birds do not, however, sing very often in the heat of summer; yet one bird is singing high up in the air above the trees. We are not able to see him, but we can hear his sweet, fleeting song. It is the wood-lark. This country is the wood-lark's paradise, and the little fellow's song rings out on the still night, and rings again. Then there is silence for a time.

A shrill complaining bleat of a cry comes from the road in front of us. A frog is in the grip of some owl. The sooner it stops, the better I shall like it, for the bird is eating the poor thing alive. Owls are numerous here; they flit about in all directions, like the bats. These farmhouses and the out-

buildings are peculiarly adapted to all their require-
ments. A bell-turret or old pigeon-cote would be
' onnateral' without its owls. The houses have their
own special pair of these birds, that usually sit in the
daytime in one of the lumber-rooms at the top of the
house, snoring and waking up, looking at each other
with one eye, to go to sleep again and snore.

I have never seen pole-traps set here to capture
the beautiful mouse-hunters. I hope that this may
never happen, though, as land even here has changed
owners, there is no saying what innovations may be
made. It is a pitiful sight to see one of these feathered
benefactors sitting with half-closed eyes, held by its
legs in a pole-trap. It makes one feel just for the
moment a strong inclination to hit out at some one,
and I have at different times given my opinion pretty
freely on the subject to those who ought to have
known better.

All those who live in the woods and fields—that
is, who get their living there—know very well what
creature or creatures do mischief. They will tell you
that rats and mice do most, and that owls make short
work with these. They begin to hunt before the sun
sets. You can tell the farms they come from, just off
the roads. One pair of owls will not poach on the
hunting-grounds of others ; each farm has its own

lands, and its own owls to hunt over them. When
the farms are a mile and a half or, in some parts, two-
miles apart, this owl question is very easily settled.
The brown owls and the long-eared owls (the title of
'long-eared' being given them because some feathers
stick up above the rest) float round the farms; the
long-eared species occasionally, the brown owl or
wood-owl frequently. If he clutches a young rabbit,
or, for that matter, a dozen, the farmer will have less
of these to nibble at his crops; but this does not
take place so very often, for the simple reason that
the doe-rabbit is a most watchful mother. Now
and again one daring youngster does not heed
mother's warning drum, and then the brown owl
grips him. When he visits the farm, you must look
for him not in the rick-yards, but where the brook
has been widened out into a shallow bay for the cattle
to drink at. If you do see him, it will be because he
is diving through the openings in the trees, or floating over the tree-tops. Rats and mice come out to
drink at night, in fact they are night wanderers; but
it is not always a rat or mouse the owl is after. He
wants a fish, and that fish is a trout. All drinking-
places for cattle are very productive of insect-life.
The trout know this, and make their way from under
the bridge, over the shallows. You can hear them

scuttle over the stones, and so can the owl. Also
you can hear them feeding. They are not large, only
the usual brook size, that of a small herring. Either
in coming or going a pair of owls will have two or
three in the course of the night—that is, if all things
are favourable, for trout are capricious. They do not
feed in one place, or confine themselves to one course
of feeding. There is plenty and to spare, for the
nature of the waters they frequent prevents their
being captured. Great rifts run through woods,
covered in by thorns and briers. Try to wade down
such waters, and find out what the stoat-flies and the
midges will do for you. It is a very old proverb,
that those who use strong language catch no fish. If
you attempt to angle here, you will prove the truth
of that old proverb to a dead certainty. Sometimes,
when the farm lads know that a brown owl has
young, they will go and turn the cupboard out, as
they say, and a very interesting sight it is. If fish
are to be got, portions of it, or the whole fish, will
generally be found there, and the barn-owl keeps a
good larder. You are not able to find it very often,
but when you do, you will find something in it.

I have seen my own owls pack their surplus food
away in the most methodical manner. After I had fed
them up until they refused to take any more, then

I have tossed half a dozen mice down in the room ; and it was fun to see them poke and push and pat to make them lie nicely, of course out of sight.

Both wind- and water-mills are in full force in this district. The mills are like the farms they grind for, very old. The water-mills, from their secluded position, are naturally the places about which to look for fowl or fish, the mill-ponds being large. You may be in such places for a whole day, and the only sounds you will hear will be the clack of the mill-wheel, the cry of waterfowl, and the splash of fish. One spot I know, that lies sheltered in a great hollow or coombe, on the very edge of the forest, which is very beautiful. A large old house, half mill half farm, stands in the trees ; they are all around it. In front is a large pool, reed and rush fringed. Great masses of weed float on its calm, deep waters. Coots, ducks, and dabchicks are going here and there in all directions ; and as to kingfishers, this spot has a local reputation for them. Other birds come that are not often seen elsewhere, and animals also. The place is very quiet, in fact out of the world, in the full sense of the term ; the old house in the trees, the hills above, and the pool in front mirroring the beautiful surroundings. Summer is the time to visit the Weald. When the fox barks, and the vixen answers him with a scream like that of a child in pain, the country wears a different aspect, and the winds are keen. At such times those long, green lanes are not to be travelled over easily, yet I have been there in summer and in winter; and even when it was snow-covered, I have found beauty in the Weald.

I FEAR that this subject is not a favourite one with the public generally ; but I must ask those who have read my articles on wild life in its other shapes to bear with me in this one, although at the first glance the title may not seem attractive.

The fact is, few are acquainted with the real nature of the beautiful creatures, as they really are, divested of all technical 'scientism,' and apart from those specimens preserved in spirits, with which all are familiar. To know them properly one must examine their haunts, where they find food, and bask in the sun ; also the spots to which they retire when cold weather sets in.

A straw hat, carefully manipulated with the thumb and forefinger of my right hand, has been my very easy method of capturing them, and, what is of vital importance, without the least injury to myself or them. Up to the present I have, fortunately, never been bitten ; quickness and some slight nerve are all that are required in picking up a viper from two feet to nearly three in length ; there must be no bungling.

The common viper or adder, *Pelias Berus*, is the only poisonous reptile which is a native of this country. The colouring of the creature, which changes according to the locality it is found in, is, in fact, as variable as that of the common brook-trout. But one unmistakable sign will single out the viper, namely, a row of connected zig-zag markings down the back from the head to the tail ; also, there is a well-defined, V-shaped, dark mark about the centre of the head— the viper's monogram. No matter what the size or colour may be, the creature that has these markings is a viper, and nothing else. The few accidents that do take place—and they are very few indeed—are caused by grown-up people, or children, picking flowers, accidentally treading on them.

These accidents occur to those who are ignorant of the creature's habits, and mostly when the primroses and violets bloom ; for that is the time when the vipers indolently sun themselves on the grassy banks warmed by the sun. It is best to look round for a few feet, before you stoop to gather your violets or primroses, although, as a rule, the creatures will get away if they can. They never attack wilfully, but when the hand touches them as they lie coiled up, exactly the colour of the dead leaves, they strike, naturally, in mere self-defence.

I have lived where these reptiles were very numerous ; yet not one man, woman, or child, have I known to get bitten, although the creatures were round about in all directions. To have a large one coiled up within a foot of you is quite close enough ; stand still if you please—that is, if you are not prejudiced against the whole family—and watch the creature. It is coiled, and the neck thrown back over half the coils, the pupils of the eye are exactly like those of a cat, all ablaze, and the tongue plays in and out, ready for instantaneous action. But finding that you do not molest it, it slowly uncoils and slips away. I have good reason to think that the vipers are very much abroad in the hot nights of midsummer ; for I have captured them crossing the roads, from the fields that bordered them, very early in the morning, in fact at sunrise. There they had been hunting for the large-headed, short-tailed field-voles, which are quite as large as any ordinary half-grown rat. I know this, for I persuaded them by a very simple process to show me what had caused their aldermanic girth. The voles had only just been swallowed. The pupil of the viper's eye when at rest shows as a dark vertical slit ;

this indicates crepuscular habits—at least it appears generally to do so.

'We're goin' to hev a middlin' spell o' hot weather yet, I knows,' was the remark made by my woodland companion, Waggle.

'How do you know that?'

'Why, them 'ere crawlers is comin' from the hills down inter the holler, fur tu git drink; they bin comin' fur this two days. I see 'em crossin' the road lots o' times yesterday. Cuss them things! they gives me the creeps. You ain't never goin there, be ye? You'll come to mischief some o' these days over them var-mints. I'm a comin' with ye, I ain't a' goin' tu back out, after tellin' ye on it, not me. But I say, old feller, if we does run up agin one, don't you hive him; that 'ere last hivin' job did fur me. Why, if anybody had offered me fifty gold suvrins I culd'nt 'a done it.'

We were walking down the dell track very unconcernedly, and in an unobservant manner, when my companion yelled:

'Look out!' and from some red-brown fallen bramble-leaves, over the toe of my shoe, shot the largest viper I have ever seen.

'Cum back, I tell ye! Damn the thing! you ain't a goin' tu tackle that; its most natur'ly sure tu be the death on ye, let go on it!'

H

I had caught it by the tail, and snatched it out of some trailing branches, jerking it past his legs.

' Are ye goin' tu do fur me as well as yerself ? There he goes, hooray ! the varmint's gone.'

' Not quite!' I cry, as I crash through the brambles and pin the viper behind the head. ' I am all right. Come and have a look at it. It is a beauty.'

' Hev ye got it by the head or the tail ? If ye hev got it by the tail, I shan't come.'

' I have it by the neck, with the body coiled round my wrist.'

' Wun't believe it ! ' he shouts, as he dashes through the underwood towards me. Then, with one look of horror, calling out ' I'm off out o' this,' he vanished.

It was a splendid specimen, twenty-six inches in length, very thick in the body, rich bronze-red in colour, with black markings, and it had a bloom on it like the bloom on a plum. When I reached the road from the hollow, I found a group of the natives waiting for me, Waggle among them ; and there I gave them a practical lesson in natural history. Shifting the viper from the right hand to the left, with a grass stem I tickled the creature's nose ; this raised its temper. To say that the mouth opened would not be enough, for the upper and lower jaws fell wide apart the two poison fangs erected, and the poison from the glands showing like very small dew-drops, as it oozed down the grooved fangs. Also I showed them the teeth behind, ready to take the place of the first, if through accident these were broken. The harmless but dreaded tongue was then examined, and other matters gone through with, which I need not detail here.

As I left them with my captive, I heard Waggle's voice: 'If you tells me agin,' said he, 'as he didn't catch that varmint fust by its tail and then by its head, you'll hev to upend yerself (fight). I tell 'ee, I see un do it.'

To settle the matter, I walked back and placed the viper on the ground; it coiled instantly. 'I will give five shillings to the man who picks him up,' said I.

'Not fur five hundred,' cried they.

I held it before them. 'Ye wunt feel niffed like when we meets ye, if we gives ye plenty o' elber-room, mister,' was their comment on this last performance.

In the course of sixteen years' observations of the reptiles in their own haunts, I have seen many varieties of the same species, of all sizes and all shades of colour. But one and all had the poison mark—the zig-zag markings. The situations they live in are as various as their shades of colouring. In the fir woods we find them, on the commons, on the sand heaths, or stony ground well sprinkled with thorn scrubs and trailing brambles. Such places are a paradise to them, and they reach a large size there. In birds' nests in the trees, where they coil themselves after eating the young ones, in the water, and at times on the door-flags of the cottages, they are seen ; and yet not one person in a thousand ever gets bitten. Their food is composed chiefly of mice ; the vipers hunt for the little thieves in the most persistent manner. Small birds, both old and young, frogs, and at times lizards, they devour ; but young mice are delicious morsels for them.

Humble-bees, the large and the small species, make their nests in the mossy ground, and mice have their homes in the same localities. The mice not only rob the bees of their honey, but they kill and eat the bees themselves. The vipers eat the mice, and so all things work well.

Should anyone be so unfortunate as to be bitten, if possible, let him go at once or send to a medical man. If none be near, tie something tightly above the part bitten, and press the punctures with the hollow of a fair sized key, then rub the part bitten with olive oil, until the doctor comes.

The poison is virulent enough to cause death in small animals, in any creature the size of a small dog, for instance ; or to a weakly child it might prove fatal, unless immediate measures were taken. Grown-up people, unless their blood were in a bad state, would suffer intense pain for a time, and probably have a fortnight's illness afterwards. All those not well acquainted with them I would advise to let them alone, and not handle them, dead or living. If those who kill them and leave them in the road, or paths, would crush the heads of the creatures under their heels, crushing them completely, they would be doing the public a service ; for the poison is as active in its action after the viper is dead, if a child should get pricked, as it is when the creature is living.

Very beautiful but dangerous creatures, that are always ready to get away if they can, they are ; and much that has been written about them is mere nonsense of the worst kind. Fortunately, the viper is the only venomous reptile we have. When the cold weather sets in, they lie up for the winter, under the dry moss and heather-tangle of the heaths ; or in old birds' nests, full of dead leaves, up trees, in faggots that have been stacked, and at the bottoms of old posts gone rotten.

This a friend of mine found out rather unpleasantly. He had dug all round a decayed post preparatory to placing a new one in its place. As he stooped down to put a cord round it, in order to lever it out of the ground, he saw something move that he at first thought was a dark ball of earth falling.

A second glance showed the ball to be a knot of half-grown vipers, just beginning to uncoil and show their heads.

' Hand me that hammer, will ye ? ' Thud. ' That's done 'em ! '

It is still a vexed question, that of their power to protect their young in time of danger by taking them into their gullets. I have never seen the action ; others have positively stated it to be a fact. They are not so large as lobworms. The female has been seen coiled up with her young on and around her ; on being disturbed, she threw herself straight out as a stick, her mouth wide open, hissing loudly, and then darted away. Although search as far as possible was made about the spot where she had been coiled, and where she straightened herself out, not one of the young ones could be found. They are probably like the young of some of the small waders, which conceal themselves so well that you may walk over them without seeing them.

All the instances of vipers that have come under my notice have been unexpected ; and in the hurry and confusion caused by the hissing and struggling of a furious reptile that one has almost trodden on, time will be gained for the young ones to go somewhere. The enraged mother has been killed ; but although, to all appearance, her offspring seemed to go down her throat, not one was found there when she was examined.

Now the question is, do the young ones glide beneath her, and hide at once in the surroundings ? Any small cracks in the sunburnt soil would be sufficient for that purpose ; this I think may be the case.

The young of vipers come into the world alive ; and although at that time they are incapable of doing harm, they will if you meet with them coil and strike like the adults ; a decided case of hereditary spitefulness on their part. No viper can strike more than half its own length. The creatures are very useful in their own domain, the wilds ; they were, doubtless, created for use in these.

There is as much difference just before they cast their old skins, and after they come out in their new ones, as there is in a man who exchanges shabby clothes for new ones. So very great is the difference, that the people give different names to the same creature. Great numbers are killed every year for the sake of their fat, with which the rustics make

their precious adder-oil. Spring is the time for that ;
for, although the creatures have been in a semi-
torpid state all the winter, they are fat when they
crawl out to bask in the first warm sunshine.
You may readily know where the oil-getters have
been at work when you pass along, for there are
the headless bodies]of the vipers hung on the twigs
of the bushes that line the banks. No forester or
forester's child ever leaves the head on a viper ; it is
cut off, as I said before, and ground into the earth
to prevent accidents. The oil is firmly believed by
them to have wonderful virtues, especially in the case
of a bite from the creature itself, but I think plain
olive oil would be quite as efficacious.

We will now pass on to a more pleasing and perfectly harmless member of the family, the common
grass snake, *Coluber natrix*. This handsome and
perfectly harmless creature, under favourable conditions, reaches the large size of from five to six feet
in length—specimens of from three to four feet long
are very common ; the larger ones are found in waste
places near the woods, where gravel or brick earth
has been dug. The hollows, after the places have
been abandoned, get filled with water, and the rough
ground covered with wild tangle ; here their food
is in abundance—namely, frogs, mice, birds and

their eggs, especially those of the latter that are ground builders. One that was killed by two of our foresters was six feet long, thick in proportion, and it had a small wild rabbit in its inside. As these secluded places are rarely visited, except by those who, like myself, go to them for the express purpose of watching the reptiles, they live in peace. Rough rails are generally put round such places to prevent the stock from getting into the deep pits of water, where earth and gravel have been dug; so that nothing disturbs them. From these places the large ones make long excursions, for purposes best known to themselves ; the one that had swallowed the little rabbit was killed a good mile from his moist haunts : he had been seen for weeks going to and fro. On my telling some natives that I would have given them half-a-crown each if they could have told me where he travelled, so that we could have caught him alive, they stared in astonishment, and one of them said, ' Surely, goodness, ye wouldn't ha' bin such a goat as to ha' tackled that, wud 'ee ? '

I replied by asking them what they had done with it ; to which they answered they ' hed twisted un roun' a stake an' showed it to their master, an' then it was tossed away.'

It is a wonderful sight to see a large snake

gliding in and among the branches of bushes and trees.    The perfect ease and rapidity with which they will glide over the slender branches must be seen to be properly understood.

At times they may be seen hanging head downwards ; and at the first glance you might think it was a dead snake flung up in the branches, so apparently lifeless does the creature hang, motionless. It is all a sham ;  it is only watching for some dormouse or bird to come near.    Break off one of the long rush-stems close at hand, and  touch the nose of the snake ; it draws its body up in a flash, and all you will see afterwards will be a shining streak shooting through and over the branches and twigs.

These creatures are  frequently represented in unnatural positions ; as, for instance, coiled round the branches of trees.    This is not correct ; they do at times coil round objects for the purpose of conceal- ment when pursued, or they fit themselves into the angles of old  masonry or brick-work in the most surprising manner ; but when they travel over the trees or bushes, they simply glide about like ani- mated whip-lashes.  Some consider reptiles, the snake family especially, to belong to the lowest order of created beings ; but this is not the case, and their structure shows that they are highly fitted for the place they hold in creation. As geological research tells us, all creatures now living on the face of the earth and in the waters, both salt and fresh, are really the dwarfed forms of mighty creatures that held sway on the earth and waters before man left any record. Tropical countries—South America for one—have huge serpents still ; but even these would look small beside the monster forms that have passed away—all we have being their skeletons or portions of them, that have been kept for ages, safe in the keeping of Mother Earth.

The general colour of the common snake is grey-green, more or less bright, dotted with black spots. At the back of the head is a yellow mark, bordered with black ; the under parts are generally yellow, marked with black. If this is borne in mind, it is easy to distinguish the snake from the viper at a glance.

Although as harmless, so far as human beings are concerned, as a worm, the snake will show fight, and coil, like its dangerous relative the viper. But it is all show ; it begins and ends there. Many times have I surprised one in the open, and held my hand down for it to strike at. Its blow is only like having a gentle tap with a button. The snake does not open its mouth when it strikes ; the tip of the muzzle only touches the hand.

One fine specimen, perfectly tame, that I had when I was in the forest lands, would rest perfectly contented hanging round my bare neck for hours as I worked, to the horror and wonder of the rustics who were about there.

At times it would examine my face with its tongue in all directions, making a most minute inspection of it.

Three or four men who were near me once chanced to see this. From a respectable distance they told me that they believed that I had 'dealings with familiar spirits an' old Cocky-Hoop,' their name for his Satanic Majesty. This opinion of me altered before I left the district; but if I visit them now, as I do at times, before they shake hands with me the old question is asked : ' Ain't got any on 'em 'bout ye now, hev ye ? '

I have, in the course of my lifetime, kept them all ; but that is a true old saying, ' A place for everything and everything in its place,' and a well-kept and well-ordered house is not the place to keep reptiles in. The line must be drawn somewhere, and the lady of my house draws the line tightly at snakes. Yet, if the general public knew the real nature of the creatures, I feel sure folks would look with interest on the very beings they have turned away from before with feelings of disgust.

The smooth snake, the Coronella, feeds on lizards, small snakes, and slow-worms, the deaf adders of the rustic population. This is our miniature Ophiophagus, or snake-eater. It is harmless ; in this respect differing from the large and deadly Hamadryad, which is really a monstrous snake-eating cobra found in South-eastern Asia. The coronella will bite, but there is nothing to fear from that. This creature is much smaller than a common snake ; it has a smooth look that the other has not ; there are other distinctions difficult to make plain] on paper. They have had specimens of it in the reptile house at the Zoological Gardens, where the difference between the two can be easily seen.

The snake is local in its habitat ; up to the present time, Dorsetshire and Hampshire are the two counties where it may be procured. Dorsetshire I am not able to say anything about, but of Hampshire I can speak.

The blind-worm, or slow-worm, *Anguis fragilis*, is not a snake ; it really belongs to the lizard tribe. If ever a poor innocent creature has been persecuted, this one has ; and, simply for untold benefits conferred on man by the beautiful, harmless being, he, as a rule, has crushed it to death whenever he has seen it. Slugs, caterpillars, and other pests are the food of the slow-worm. Not even the viper is held in such abhorrence by them as the innocent slow-worm.

If I have not been near enough to pick it up and put it in my pocket, when, in the company of country friends, I have seen one lying on the road, no mercy has been shown ; crunch has gone the heel of a heavy boot, and it was done for. If I have been able to save the harmless creature in the way mentioned, I have had all the road to myself afterwards, with the echo of various opinions on my conduct in general, couched in the most forcible language they were capable of using, sounding in my ears.

I have seen fine specimens of this really legless lizard. Their colour varies from a warm grey to a light bronze.

The title of blind-worm is wrong all ways, for the creature has very bright eyes, and also eye-lids. It is so very smooth you might easily pass it by, taking it for the broken handle of some old red-glazed tea-pot, as it lies curled on the ground.

I sincerely trust that the poor creature will not be persecuted as it has been. So far as my own influence in the matter goes, in some instances I have been successful in persuading a few to spare it; but only some, not all, for the prejudices of tradition are not easily put aside.

The handsome sand-lizard, *Lacerta agilis*, comes next; this is larger than the common heath-lizard, which is below on our list. Its general colour is a sandy brown, clouded with a darker brown, having rows of black spots with a white dot in the centre; but in this species, as in those others we have described, the colour varies greatly, according to season and locality. Some are so green in tone that they have been taken for the green Jersey lizard, which has never been found in England; it is not indigenous, at least. This is also local to some extent; but dry heaths and grass-covered banks, bordering the sandy tracts, are not the only places to look for it, for we have seen it basking close to the edge of a salt marsh beside the tide, in the sweltering marsh-harvest times of past years.

The green, or Jersey lizard has been imported and turned loose in suitable localities by a gentleman living near Dorking. But the creatures have not prospered—they gradually dwindled away. Solitary specimens have been seen and killed, not captured. From this circumstance the report spread that the green lizard had recently been discovered to be a native of the southern part of England. This is how energetic compilers make stock.

I have seen golden-green lizards, sports or varieties of the sand-lizard ; but it must be remembered that there are many shades of green. The green lizard proper is a large distinct species, and not a native of any part of England.

I have given the scientific titles of the creatures in this article in order that there may not be the least trouble in identifying them in any scientific collec- tion. Recently it has been the fashion to find and describe fresh species, so called, and this has created confusion ; varieties, or sports of nature, are not species, although some very erroneous so-called new ones have been formed from them in books.

Those whose business leads them to pass their lives in London and other centres of commercial life have, as a rule, little time for studying natural life, if they have the wish to do so. To these the magnificent collections of natural history now scattered about the country, broad-cast we might almost say, must be a boon indeed.

The common lizard, heath-lizard, or nimble lizard, *Zootoca vivipara*, is very common, as one of its titles plainly denotes. Its general tone of colouring is a golden brown of various shades—all this depends on season—dotted over with dark spots. Some of these I have seen were like gold bronze. One of my young friends, who, I fear, gives me credit for knowing far more than I do, raps at the door at all times, day and night, up to ten o'clock p.m., to show me his captures, and also to ask me to tell him what they are. One day he brought a fine heath-lizard, telling me in the most matter-of-fact manner that he was going to tame it. The boy did tame it, and at his earnest request I went to see his nimble little pet; he had it on his hand out of doors as well as when at home. I have often seen and admired it since. It caught its own food with wonderful quickness, indoors and out, to the admiration of all who saw it, but the back of the boy's hand was the little creature's favourite resting place. It was killed by accident, the almost universal fate of unusual pets.

The young of the common heath-lizard are produced alive; so are the young of the slow-worm. The sand-lizard lays its eggs in the sand, where the sun's rays hatch them out. This is one of the mysteries that make useless all conjecture why creatures so closely allied should produce their young in different ways. Take, for instance, the first two reptiles described in this article, the viper and the

I

snake ; the young of the first come into the world very much alive, the second deposits a long chain of eggs in some heap of decayed vegetable matter, and these are hatched out by the heat generated there. The reason for this difference we really know nothing about.

The common frog, *Rana aquatica,* we have seen in great variety, of all shades and all sizes, in the water and out of it, near to water, and again a long way from it.

As there are all kinds and conditions of men, so there are all sorts of frogs ; even in common frogs there is as much difference as there is in common humanity. Lean frogs we find, also plump frogs, dirty frogs, and bright handsome frogs ; weather affects them greatly, in the same way that it does some of us at times. I remember when the frog family had a rough time of it at the hands of all school-boys ; but that is a thing of the past, at least, I hope so. He is very much appreciated by animals, birds, fish, and reptiles, so that his race need be a numerous one. Man himself pays him some attention, for he eats his hind legs as a choice morsel. The edible frog proper, which we shall notice presently, does not always supply the spring-chicken flavoured hind-quarters.

It is a most amusing sight to see a lot of domestic ducks hunting for frogs in the shallow pools, before these actually emerge from the mud where they have buried themselves all the winter. As the water gets warmed by the sun, they gradually work up to the top until only a very thin layer of mud is above them, a mere crust in fact. The ducks know all about this ; they swim in the shallow water, rake the frogs out with their paddles, and nip them as they try to escape. The quantity of embalmed frog I have seen them waddle off with has been something surprising.

The edible frog, *Rana esculenta*, is now a native of this country, but whether it has always been native to our island, or if it was introduced from the Continent in past times by the monks, I am not able to say. One thing is certain, he is to be found in one of the Cambridgeshire fens ; at least, he was found there recently. This frog is a valuable creature as a food-supply ; but it will be some time, I fear, in this country, before it is seen on the tables of the public at large. For many years I have known that the hind legs of our common frog were eaten in some counties ; but this may have been by people who had lived on the Continent, or who were immigrants into those counties.

Of the changes from the tadpole into the perfect frog I do not intend to write ; there are very few people who have not seen that for themselves.

Our common toad, *Bufo vulgaris*, reaches a large size, and I have seen some specimens lately that must be the giants of their race. The toad and the frog are great helps in destroying insect life, whether it be flying, running, or crawling, the toad especially.

If one is killed after he has been out for a night's hunting, the quantity and variety found in the creature's stomach will be something to wonder at. Bees are a special delicacy to him, and bee-keepers show scant mercy if they find the toad near their hives. In looking over some beautiful and very accurate drawings executed by a Japanese artist, I noticed one drawing particularly, which represented a couple of toads as bee-catching. If they keep from bee-stealing all is well, as they are most useful in a garden for clearing off insect pests, but their presence near bee-hives is not desirable.

The creature can be made very familiar, and it certainly has the bump of locality very fully developed. One that lived in my wife's flower borders used to come out in the evening to the front door-step, and have a look at us. After a time, thinking some one might open the gate in the dusk and crush him as he was walking about, I moved him out on to the common. The next evening he was inside the garden plot as usual. Once more I carried him away, and this time to a much longer distance. He found his way back again, however, and after that he remained with us as long as he pleased. If toads make up their minds that they like a place, they will stop there, unless you kill them, and I certainly could not do that. In most gardens now, you will see toads in the melon and cucumber pits, and in the houses where grapes are grown, for one of the gardener's best friends is our common toad.

The natterjack toad, *Bufo calamita*, may be distinguished from the common toad by a bright buff line down the middle of its back. It is more active in its movements than its far more common relative, also more local in its distribution. In other respects it resembles its common representative.

The newts, or, as they are called in the country, efts, and far more frequently effuts, will close the list of our British reptiles. The great water-newt, *Triton cristatus*, is, as its title implies, the largest member of this family, and the handsomest. Its length is from six to seven and sometimes eight inches. In the breeding season the male is dark yellow-brown in colour, with dark roundish spots; the under parts are

deep orange, spotted with black.  The sides are dotted
with white, and the sides of the tail are of pearl white.
A deep-notched crest is on the creature's back, in-
dented like a cock's comb ; this, as the creature floats
near the surface of the water, with all its legs spread
out, is a conspicuous ornament.  It is a voracious
creature, feeding on tadpoles, small frogs just after
they have ceased to be tadpoles, also on worms ; a
worm it is impossible for him to resist.

' I say, Mister, do ye want any o' them 'ere fin-
backed effuts, big uns ?' was the question addressed
to me from a couple of woodland youngsters, attired
in suits of clothing very much out of repair, ' 'cos if
ye do, jest cum along o' us down tu old Bitter's hoss-
pond, in the medder ; we'll pull ye out some quick.'

As the invitation from the youngsters to invest in
effuts was so heartily given, I accepted it.  The prospect on their part of a small silver coin closing the
performance of fishing for effuts, may have had a little
to do with their readiness to oblige.

On reaching ' old Bitter's hoss-pond,' the proceed-
ings were of the simplest kind.  Each lad cut a long
slight rod from the sallow-bushes, tied a piece of stout
thread to it, and then to the end of the thread they
tied a worm by the middle.  As the large specimens
showed in the clear water, the worm was dropped

down in front of these. They at once fixed on to it like a bull-dog, and so were jerked out on the grass. As I stooped down to pick the first one up, a very fine one, alarmed shouts proceeded from the pair of shavelings.

' Doan't touch un, fur marcy sake, afore we settle un fur 'ee ; whativer they bites swells till it busts.'

It was not the least use reasoning, it was only time wasted ; not even the sixpence given them quieted their fears on my behalf. ' Sumthin' wud cum on it, sum time or other,' they said.

The children reasoned, although not correctly, from inferences drawn from their own observations made in the forest. ' Now look here, Mister, afore you gits pisined. They must be rank pisin, fur their bellies is 'zactly the same colour as them 'ere musheroon things what grows under the fir trees, an' they'll settle ye.'

The newt we have described leaves the water and hides on land, as well as under the mud of the ponds it frequents. After the breeding season is over it is a very different creature. Its crest is gone, and all its bright colours have faded to dingy shades, so that it does not look like the same reptile. The common smooth newt, *Lissotriton punctatus*, is to be found almost everywhere, if the water is clear. It is preyed on by its relative, the great water-newt, and by fish. This newt varies much in colour; the general tone is brown-grey above, bright orange below, marked all over with dark spots, and the crest is very often tipped with red.

There have been two more kinds of newts figured and described in works on reptiles. Whether they are species, or varieties, is of little consequence. I have seen so many strange changes, influenced by season and locality, in the creatures I have briefly, and I trust plainly, described from the life. But after long study, I should still hesitate before I gave my opinion from any scientific point of view. I could have given the dry technical details of all those I have written about, but have not done so, lest my usually indulgent readers should skip the article; what I have written, however, has been from personal knowledge of facts that are beyond dispute.

In conclusion, I must say a few words about the real use of the creatures. The stronghold of reptile life in the country, as in other lands, has been the swamps and wastes. The fens and marshes have recently almost passed away; fertile corn-fields having taken the place of sedges and reeds. In the hot, reeking fen and marsh-land summers of my younger days, snakes and vipers swarmed. When the frogs spawned,

.and the snakes left the sedge-beds to come into the dykes to feed on them, it was a sight beyond the belief of all inland dwellers. There the snakes were in numbers. They did not have all their own way in the matter, moreover, for birds of prey fed upon them in their turn. Kites, the forked-tailed kites, harriers, hooded crows, those that remained to breed with the carrion crows; herons and bitterns came also; but the frogs formed the principal food for all the others that had their homes in such places—otters, polecats, stoats, weasels, spoonbills, nearly all the duck tribe capable of swallowing them, besides the fish, pike, perch, and eels.

It is all very well to state that such and such authorities give certain lists of creatures as the food of others. Most of this is sheer humbug ; only those who have lived with the creatures in their haunts, crawled like the reptiles around them, waded in foul marsh-water up to the neck, and got the deadly marsh fever through it, know really much about that.

I have seen a duck knock and bang about a small reptile in a most determined manner, and then swallow it down like a worm, after it had been quite disabled. No reptile from ten inches to twelve inches is any trouble to a duck, after it has been killed and flattened out by being rapidly passed through the birds serrated mandibles. In this manner, in past years, the so-called balance of nature was kept up—one creature formed the food of another. Those creatures even that were most loathsome—or at least so some of them would be considered—were actually necessary, through their destruction of other creatures, to make those ague and fever-stricken districts habitable for human beings. When the reptiles ceased to be, the creatures that had fed on them were seen no more, which speaks for itself. Independently of this, another cause at times cleared those reptile-haunted flats, for, when at un-usually high tides and in rough weather the salt water broke the banks and drowned the flats with a salt flood, 'it pickled 'em,' as the marsh folks said. If the creatures that preyed on the reptiles had not been wantonly killed off in the southern counties, the latter would not be so numerous as they now are in some districts that I visit.

## MORE ABOUT THE OTTER

'I WAS looking out for you to pass ; come in and see what I have to show you.'

So spake one of my friends. Going to an out-house, he brought from it a fine dog otter, dead, of course. 'Look him over at your leisure,' added he. This I at once proceeded to do, with mixed feelings of pleasure and regret ; glad to be able to examine once more, minutely, one of the most perfect of animal mechanisms ever fashioned by the great Creator, but grieved that this brave, beautiful creature should have been shot.

I may as well at once state that my opportunities for studying the otter have been few and far between. I have, perhaps, only seen the animal five times in the whole course of my life, fully and properly that is, though I have been looking out for him for over forty years. But I can assure my readers that these opportunities have been made the very most of. A born naturalist is always on the alert. A few disjointed remarks heard casually, a sign noted here or there in the creature's haunts, are all that is needed to bring one as quickly as circumstances will allow to the spot. No matter if it be night or day; if the animal is there he will see it. Short though the glance may be, quick eyes see and gather much, and one comes away well content, even though one has been bogged up to the knees, and bitten to such an extent by poisonous little midges that the eyes are puffed up and the tip of one's nose feels as if a small strawberry were fixed on it. These are trifles; the main thing is, a man has seen what 'he went out for to see.'

As I have said before, all wild creatures will, if permitted, draw near to man and the roads that lead to his dwellings. The otter is not an exception to this rule. When he has been domesticated, he requires a large amount of liberty. Those that have been best kept have shared the kennel with sporting dogs. They knew their own place in it and could keep it well, and there they have been strong and glossy of fur. Otters vary very much in size and colour, some being comparatively light in fur, others darker. Affectionate and intelligent by nature, they will attach themselves to man when well treated, and will gambol round him like kittens. You must not, however, touch the head of any otter that is domesticated. There have been exceptions to this, but it is a wise precaution to observe. Animals have their little whims, like ourselves.

To return to my friend and his dead otter. I have the latter at my feet, and there place him in one of his favourite attitudes, that of floating calmly with the current down some stream, or over the sharply running river-shallows. Head, body, and the long, powerful tail are in a straight line, his body flattened out, so that his webbed feet act like four short oars. This is a characteristic position, but life is wanting. A while ago his grey-brown fur glistened in the light, falling in so closely with the shadows under the banks and the wet mud of the river-side, that he looked more like a wraith or water-shadow than the strong, active creature he really was.

As I look at him some old memories come up, and I am young once more, standing on the sloping boards of an old weir. The leaves are falling in showers, like faded flowers blown all about, for it is November.

A white hoar-frost is on the weir boards and on the grass in the meadows, but that does not prevent me from plunging into ten feet of water so as to swim underneath the overhanging bank in search of the curious seal that those partially webbed feet have left behind on the soft sand and mud.   I would plunge in there again if I had the chance.   I do not feel too old for that yet, but the place he still frequents has changed owners, and all do not enter into the spirit of practical natural history.

Here is another haunt of the otter—a reach of sluggish river-water flowing through meadows that are almost level with the river itself.   On one side are high banks, copse growth, and fine old trees: some of these have fallen.   On their grey trunks the yaffle shins about, yells, laughs, and yikes to his heart's content.   Now and again he pokes his head over the side of a limb as he clings to the bark, makes a dive off to the next tree, taps, and peeps again.   Here, also, come the greater and lesser spotted woodpeckers to play their side-drum solos in their playing time. Here, too, in their season, great masses of cream-coloured meadow-sweet spring from the lush-grass, small thickets of it scenting the air far and wide with their fragance.   Mingled with these are clumps of purple loosestrife; the warm brown stems of the meadow-sweet and its rich creamy blossoms harmonising well with the rich colouring of the loose-strife and the golden-green grass.   In spring the lilac spots of the cuckoo-flower—the lady's smock—are in patches at intervals ; great rushes, yellow king-cups, and irises too.

This is the home of the moor-hen, but not of the otter, only his haunt. All amphibious animals seek the land, either to rear their young or for the purpose of procuring food. If their mode of life compels them to be in or on the water, or paddling about the margins, it is imperatively necessary that they should have warm and dry sleeping and resting places. Very adverse circumstances will drown the otter, the water-rat, and that little diver the dabchick, if they do not clear out in time. When I have threshed the branches of a low oak with my ash staff, to find out what creatures had sought shelter there, I have known water-rats plunge from boughs seven or eight feet above my head into the water below. At times they are flooded out like ourselves ; yet they must live, and it is a matter for constant wonder, the way they can adapt themselves to circumstances.

Our otter is an adept at this. I have been by the river-side all the morning without having my float once moved. That does not trouble me, for plenty of busy life is to be seen in this particular spot. Birds and insects are enjoying themselves to the full. Presently the cause of the fish not biting is explained. As my eyes rest on the far side of the water, I note what looks at first sight like a bit of the low bank crumbling before falling into the water. A second look tells me, however, that it is something different ; earth falls down into the water, whereas this supposed clod of earth has tumbled up so to speak, on to the meadow. So closely did his fur fall in with the tone of the bank that if he had remained there stretched out he would certainly have escaped notice. He is only a very few moments in sight and, as is too often the case when something worth seeing is to the front, I have left my glasses at home. An instinctive dive of the hands to both pockets, and words are muttered that would not look pretty in print. Light flashes from his fur catch my eye as he moves through the meadow in a direc- tion that is very familiar to me ; then I lose sight of him. He is making for a faggot-stack belonging to one of the cottagers on the estate—one whose cottage is unpleasantly near the river at flood times. The stacks of faggots and cord-wood, as a rule, are at some distance from the dwellings. Cord-wood is the smaller limbs of oak, the lop and top of the branches when the trees are felled. The largest pieces are placed on the ground crossways ; on these the faggots are laid lengthways, and upon this foundation the stack is built. Then it is very roughly thatched with spray stuff. Air can get through it and it is soon in first-rate condition for burning, being strong and clean.

When flooded out of his home last winter, driven by the swirling torrent, he was making his way to the copse, saw the stack was high and dry, and went under it to find one or two rabbits comfortably settled there, snug and warm. As he does not feed on fish alone, he killed the rabbits, and so provided food for himself to last a considerable time. Cottagers rarely keep dogs; in fact, they are not allowed to do so on large estates unless they are shepherds. So no dog was there to whine and scratch round about our otter's hold. The spot is away from the game-covers, too, so he is safe—a fact the intelligent old fellow is perfectly well aware of. Only those who have a real regard for wild creatures will find him out, and they can be very safely relied on for not speaking about him to those who would do him harm.

All manner of excuses, more or less plausible, are made for killing the few wild animals that are left to us. The otter and his relative the badger, 'gallant beasts' as they have been called by those who are most intimately acquainted with them, can never be classed among vermin. The poor creatures are simply trapped or shot to be set up in glass cases. 'He has just been killed, been got out of the drain; here

K

you can look him over quick ; master says I've got
to put him in a basket and to send him off for stuff-
ing.' Before the local boards for sanitary measures
existed, gentlemen's houses near the river had large
drains leading from the kitchen buildings, the brewery
included—and, like old Simon the Cellarer, many
were brewing all the year round—down to the river.
These drains only led from the offices mentioned.
As nearly all of them were three feet in circumfer-
ence, having a good fall, they were always clean ; in
fact, it would have been difficult for small matters to
have stopped there. Refuse food from dish-washing
was continually coming down into the river ; this the
fish found out, and consequently they were wont to
gather there in numbers—good-sized ones. The otter
found the fish out at the drain mouth ; and when the
river was low he ventured up the drain. It was warm
there, and, moreover, a lot of plump frogs regaled
themselves in it. After sampling a few of these he
ventured further, until he saw daylight. Places were
not barred up in those days as they are now. He
just poked his head out, then turned round and
scuttled down the drain into the river, much faster
than he went up. One day he went up and did not
return. A fowl or two had been missed—these
creatures have a habit of gallivanting 'round such

places. Our otter was shot in the very act of secur-
ing a bird for himself.

This sort of thing he is guilty of now and again,
not often ; he is driven to it by force of circumstances, for which due allowance ought to be made.
No such thing as an unmixed blessing has come under
my notice as yet ; and that the otter is in his proper
sphere and beneficial to the waters he frequents, I have
not the least doubt. Of course, I am not writing
from the trout-preserver's point of view, but simply as
a naturalist.

Various local arrangements, and I believe notably
the food supply, affect the variety in the colour of the
otter's fur. Some, I am aware, hardly agree with
me as to this. It is a very easy matter for him to
get a rabbit at certain seasons. I have drifted down
a narrow woodland river, with high banks on either
side, in hard, frosty weather, when a fringe of ragged
ice was hanging to the edge of the shore, and seen the
rabbits sitting close to the edge half asleep. They
did not move, although I was not more than two
yards away from them. Nothing more easy than for
the otter to glide down the bank right down on to the
rabbit and to grip him. There is no reasoning with
an empty stomach. The moor-hen at such times
clucks and paddles a few yards away from the water's

edge, picks at a bit of dead sedge just out of the water, flirts up her tail, clucks, and is returning to the bank again. She never reaches it, for she goes under the water most mysteriously—seized, in fact, from below by the otter.

Rabbits and rats tunnel through the river banks, right down to the edge of the water ; and rabbits can well be spared for the otter when he needs them. That is my opinion of the matter at least.

One more sketch of our friend before leaving him. I am standing idly by the side of a deep and narrow lagoon, formed by a river flowing up over a water meadow. A narrow inlet direct from the river sup-plied this bit of deep water. Three feet at the most in width the inlet is. At one end of this small lagoon water-lilies and bullrushes flourish luxuriantly. Good pike and eels are known to have their home here, but for some reason or other I have always failed to catch one. All at once, as I stand there, one of the lily-leaves dips under the water. I fancy a water-rat has gone over it ; but other leaves follow, the long stems writhe and sway to and fro, and the yellow flowers dip and nod. Something invisible causes a splashing on the surface ; then a moor-hen squatters up, and runs over the meadows, clucking loudly, evidently flustered by the commotion below. The leaves are agitated more violently than before, and a rippling furrow, low down, shoots towards the narrow inlet. The furrow breaks, up shoots the head of a pike, and beneath it a brown head shows, fast gripped on to the other. All is done much more quickly than I can write it.

As I walked away I heard a shot, and, passing on the opposite bank, was hailed by a man of my acquaintance, who shouted to me that he 'had got him.' By the tail he held the otter. Its love for pike had cost the brave creature its life.

## SMALL DEER—OUR RODENTS

No animal has had more enemies, perhaps, amongst all sections of society, than the common brown rat. Farmers, gamekeepers, sportsmen, ratcatchers, soldiers, sailors, and sewermen, and last, but by no means least, our household cooks—all have their grievance with regard to him, and he suffers under their undying persecution.

Here he is, however, and like the typical Jew he seems to flourish in spite of all, and to be about as universally distributed. The hand of every man, woman, and child is against him, and added to these he has his natural enemies in his own animal world, who, if man would only leave them to their appointed task of thinning the ranks of these depredators, would do it to perfection. Unfortunately, man supposes himself to be a wiser orderer of creation than He who called all creatures into being, and so the balance of nature has been upset.

A strange mixture of courage, caution, audacity,

and perseverance ; all these qualities being brought into play by him as the moment demands. I have never yet seen either a poorly conditioned rat or a lean sparrow ; both these creatures have the wisdom and the pluck to attach themselves closely to man and his fortunes. The rat is so intelligent, too, that were it not for that unlucky long tail of his, he would be made a domestic pet of more frequently than he is. That very useful appendage tells against him, however.

One thing can certainly be said in his favour, and that is, that he acts, with a host of other creatures, some of them equally depised with himself, the part of scavenger. A first-class scavenger, also, he is, and a good swimmer ; he can climb, too, anywhere. I like to see him best out in the country, to which he resorts—as the favoured classes of townsfolk do—in the latter part of the summer. Hosts of rats quit the towns and villages for an annual holiday at that time, coming back when their season is over to the parts more inhabited by humans. All who go do not return, for, being most expert and confirmed egg and chicken stealers, they receive no mercy at the hands of the rustics. The very choicest products of the country they have a knack of appropriating. In the course of a recent twelve miles walk my way led down a hillside, on the outskirts of some covers.

There were only a few thorn trees growing there,
with now and again a clump or two of furze, some
thin ash trees, and low tangle.    As I find it a good
rule never to get over a stile before making sure
what is on the other side, I looked over and saw
two magpies at work, before their keen eyes had discovered me.    They ranged that hill from top to
bottom, one flying over and the other keeping watch,
alternating in these duties.    The thorns and furze
clumps were minutely inspected ; with their nimble
movements it did not take long to range the hillside.
A magpie will kill a half-grown rat if he has the
chance, a creature that would do infinitely more
mischief than himself.    I have known a rat or rats
take a dozen eggs from a wild duck's nest and bury
them in the soft, peaty bottom of a moorland runnel,
close to the nest.    I traced the whole proceeding, and
dug the eggs out with my fingers.    They were
deposited at distances ranging from a foot up to
eighteen inches apart.    The work had been done in
the most skilful manner, for the down that had covered
the eggs so beautifully in the nest remained undisturbed.    There was a faint trace where the eggs had
been pushed forward by the creature's breast up to the
edge of the runnel, and there were the prints of feet
in the soft peaty soil.    The arrangement of the eggs

had been accomplished after the whole had been taken. A cruel foe to young water-fowl is the rat ; a far more formidable one than any not familiar with his ways could imagine. I have seen him stalk the birds through the herbage like a cat would ; whether his quarry is in the water or out it matters not to so accomplished a swimmer and diver.

If there happens to be a kennel of sporting dogs kept near their country haunts, all the rats for miles round will congregate there to share the food of the dogs. Their impudence is amazing ; they seem to fear neither ferrets, terriers, nor traps, and the cry is 'still they come.' Heartily cursed by all, they thrive more and more.

The rat certainly provides sport for boys and men ; it is curious how if the former are seen making for certain localities, with terriers at heel, men are sure to follow ; and when the fun grows warm all are boys together. At the war cry 'A rat ! A rat !' both men and boys become as excited as if their lives depended on their killing the long-tailed freebooters.

But this is getting altered like everything else, and in some places rat-hunting is not now viewed with any favour. Formerly, amateur ratcatchers were received with open arms ; now they are regarded somewhat suspiciously. A farmer told two of my free

and independent ratcatching friends lately, in no
pleasant terms, that he did not want his rats caught.
The once admired song ' The ratcatcher's pretty little
daughter ' is also a thing of the past, sharing the fate
of another once familiar to some of us, ' Her roses
and lilies all turned to tan, When she fell in love
with the dogs'-meat man.'

That much more pleasing animal, the black rat,
is nearly exterminated, his species having given way
to the far more powerful and ferocious brown rat.
Now and again a solitary specimen is recorded, one
of the last remnants of the so-called old-English
black rat; though, from what I have been able to
gather, he is no more English than French or Irish.
Rats can accommodate themselves to any circumstance and every climate.   They traverse the globe,
paying nothing for their passage, and they land at
our very antipodes without ceremony, to increase and
multiply exceedingly in the land of their adoption,
verifying truly the adage that ' Vermin breed fast.'

In the West Indies, notably, the brown rat has
located himself in the fields of sugar-cane ; but so
also has that deadly foe of his, that poisonous reptile,
frequently of large size, called the fer-de-lance—the
horrible craspedo-cephalus—that feeds on rats.   If it
only killed these, all would go well ; unfortunately, it

also bites the cane-cutters, and when one of the latter
has been bitten, 'Quashy' does not play the banjo any
more.

'Ye little vulgar mouse,' as old Topsel calls him,
may be dismissed with a few words. Though a very
pretty creature, he is so mischievous inside a house in
various ways, that in numbers mice become a perfect
pest ; in fact, no mistress of a home will suffer the
presence of a mouse in it if she can possibly help it.
The gentlest beings I have ever known will act in a
somewhat vicious manner towards rats and mice.
And yet it is truly wonderful to what a point the
education of a mouse has been carried. I have often
wondered where mouse university is located.

The water-vole, or, as it is more commonly but
improperly called, the water-rat, is a happy, innocent
creature, that has its habitat by the water-side,
gaining its living thereby. I look on him at times as
a miniature beaver ; whenever I have seen him his
actions have been so very beaver-like. Many an
hour have I watched him by the edges of various
waters. Go where the angler will, he will see three
different creatures at varying times, close to him
throughout the day—the water-vole, the moor-hen,
and the kingfisher. The latter beautiful bird will
even settle on one's rod now and again, and that is

surely close quarters enough for observation. As to the voles, I have had them at my very feet. It is necessary, however, to swim if you wish to see some of their playing places. This I have often done, and I once had a very unpleasant experience connected with that form of research. Having turned over on my back in order the better to see something that was above me, I struck out with my legs, so shifting myself without altering my position. Under the bank I shot in among submerged tree roots, and into darkness, until the soil above touched the tip of my nose. I got out again somehow, dressed, and went home.

Some days elapsed before I fully realised all the features of that lucky escape from what would have been certain death if I had hung, caught in that network of roots. It gives me the creeps now when I think of it. The pursuit of wild things does not lead on in smooth ways by any means. One goes also alone as a rule, and I have had some very near 'squeaks' at times, and shall have again no doubt, in spite of advancing years.

I have never yet seen the water-vole eat any animal substance; though I will not assert that the creature never does this. As a rule, little Dot Beaver feeds only on vegetable productions, so far as my personal knowledge goes. You may watch him cut his provender and swim with it in his mouth towards his burrow. I have quietly drawn back at times, lest he should be startled and let go his mouthful, for I admire the industrious little chap; and one of the prettiest sights I have ever seen was his cutting the seed-vessel of a water-lily from its stem, and sitting on one of the broad leaves to eat it, holding it in his forefeet in the same manner as a squirrel would. Just above in the stream was a good dace swim, so that the current, as it ran into the lily pond, swayed the leaves; in fact, the leaf our vole rested on was gently moved first on one side and then on the other, to the full stretch of its pliant leaf stem. It was a small but very beautiful little picture; and one which I hope some day to place on canvas; the warm brown fur was brought out well by the waxen green of the leaf he sat on, which showed its dull crimson side as it was moved by the current from below. A few flower buds not yet opened, with their leaves, and just a peep of a bend in the river, made the thing perfect in its way.

I know a pool, having a sweep of flags and rushes at one end of it, which is a favourite haunt of our little friend; fine oaks line both sides of this long, deep home of the pike. When the youngsters are able to get about, it is most amusing to see how closely they are kept to the very shallowest parts.  Here they receive their first lessons in aquatic life.  Any little hollow where the tree-roots form a sort of cage for them is frequented ; any shallow where a dead limb has fallen and half rotted as it lay, will give another opportunity for them to exercise themselves in their various movements.  For the young of all creatures would, like those of so-called humans, come to dire misfortune if they were not properly looked after.  The wild instincts of all domesticated animals, also, are called into play again when the maternal passions are roused, and the exigencies of the case demand it.  When you see the voles swim from one side of the pool to the other, you can keep your rod in its case.  For depend upon it the pike are then in the sedge and rush roots, or at least just outside of them ; and the young of birds as well as the voles are out.  This is in the lush-green summer-time ; but autumn follows, and the rushes and sedges are brown, and the foliage of the oak has become a rich tawny orange in colour,  As the wind rustles through the branches the acorns fall from their cups, points downwards, with a hissing swish, swish, swish, like so many conical rifle-bullets.  These go to the bottom of the pool, where they lie hidden from view, but not lost, for wild ducks will come and dive for them, after they have picked up all they can find on dry land. The acorns all fall on the edges of the pool, so that the ducks can easily get them without too much exertion on their part ; the thicker ends stick straight up in the soft leaf-ooze.

But when the acorns fall, and a breeze curls the surface of the pool, our little swimmer and diver, the water-vole, keeps inshore ; for at that time a strip from a white pocket handkerchief on a pike trace will land one or two fish. Not many yards would our vole go before there would be a vicious snap from below, and he would be dead and crushed as flat as a pancake in less time than it takes to write about it.    In fact, my pretty, harmless friend has many enemies both above and below. The heron is one of these ; he lets drive at him on the back of the head with that bayonet-like bill of his, and then swallows him whole. Stoats, weasels, and the now rare polecat, all interview him when they get the chance, not to speak of the pike and the otter.    I have good reasons for believing that the water-vole forms part of the otter's prey at times, just as the kestrel or windfanner—a true falcon—is chased, killed and eaten by the peregrine, one of the falcon princes ; which proves that the old saying that 'hawks do not pick out other hawks' eyes' will not always hold good. But in spite of all drawbacks, this vole prospers and is merry ; man has no grudge against him—not in the southern counties at least. It is different in the fen lands, where he makes holes in the dyke-banks. The harm he unwittingly works here has, of course, to be stopped as speedily as may be, for a broken bank means drowned lands and cattle.

One of the anglers' most entertaining companions would be missing if the water-vole were not by the river's side, for fish will not always bite. Happily, other creatures have their habitat in and near the water. The black water-vole is a beautiful species, or it may be he is only a well-marked variety of the common water-rat so called. This is a smaller animal than its far commoner relative, a little fellow that looks like a velvet ball as it sits bundled up. He frequents the sides of clear, slow-moving brooks. The last one I saw was sitting on some blossoms of meadow-sweet and loosestrife, which had been trampled down by cattle coming to drink. Doubts have been expressed by some I know as to the existence of these creatures in a locality 'within an hour of London town,' but there they are for all that, and within ten minutes' walk of my own door ; and there they will be safe from all collectors, or those who pose as such. I always have done my utmost to protect small deer when they are not injurious to others in any way, in order that if I or a brother field naturalist wish to see it in its own home, it may be there to be looked at. This little velvety-ball variety of the water-vole is scarce, even in the undisturbed haunt I know of, and it would be very easy to exterminate it. I found it out after long and patient waiting and watching.

From the latter species it is an easy drop to the two species of water-shrews, for they have their habitat in the same quiet places as the black voles ; only these are on a much smaller scale, as they frequent the runnels and little dykes that run from the clear but sluggish brooks. The brooks are nothing to look at, having at intervals small brick drains where the cart tracks went over. I like them, though, for there are associations connected with them relating to a time when I could range there freely, whereas I can only visit them now by permission.

A few days ago I was standing looking over an old bridge, when an old labourer, who used to work on an estate I often roamed over, came up and said : 'What are ye sidderin' on now ? Allus up to summut ; ye looks full on it at times.'

'Why, things are so altered,' I replied ; 'one must go so far to see so little.'

'And what do ye want now—owlets ?'

L

' No, some bubble-mice I am looking out for.'

' Well, they ain't gone yet ; you go up the meadow, and if ye looks over the foot-bridge where the old master tumbled in that there day, if ye looks over deedy (carefully) like, you'll see some on 'em.

' Massy sakes! do ye 'member that there time ? It was a most 'menjous wet and desprit hayin' time,' an' the crap so thick too.   Day arter day we hed some on it ready fur cartin', an' then 'twould come down agin as ef it hadn't rained for months ; a real losin' game it was, so much on it lay fairly rotten in the water-meadows.   Master didn't say one word a-grumblin' on it to me, but we worked as hard as iver we could ; he was a rum 'un, desprit self-willed an' high-handed.   One arternoon it rained wuss than ever, an' he looks roun' dreadful, as ef he'd kill the fust thing in his way, and he hollers out, " Shut in, go to the stables, an' leave the lot on it to rot ! "   I was just behind him, an' I sees him go to one o' the haycocks, shove his hand into the middle, pull out a bit an' put it in his pocket.   Then he hollers out, " I *will* have *one* bit of dry hay out of this crop ; if I don't I'm damned ! "   An' then he rushed on to the bridge.   His foot slipped, an' he went head fust into the sheep-dippin' hole, right under.   I dunno' what a divil looks like, I niver see one ; but I should reckin he'd look

summut like the old master did arter he cum out o'
that 'ere sheep-dippin' hole.'

The modern system of drainage, although it con-
fers untold benefits on all classes of society, is destruction to the water-shrews; they can only exist under
certain conditions; and it is only on and about more
or less decayed vegetation, as a rule, that their food
can be found. It is true, I have found them in the
pure rills of a moorland district, and have watched
them nose and poke all along the margins, shoot over
the sandy bottoms, come out to run over the rootlets,
plunge in again to poke about as before; they must
have the minute forms of insect life, either mature or
immature, to feed on. But the proper place to look for
them is where the water dribbles from some old dis-
used culvert or drain in the water-meadows, where
a small pool of a few gallons of water has formed.
The spot 'where the old master pitched in' is larger
than what I have described, for although the sheep-
dipping hole has been filled up since the course of
the water was altered to make a new fish-pond, there
is enough left still to fill an ordinary water-butt, just
where it trickles down from the old drain. There is
a continual trickle, the run of some small springs.
The brick wall is going to decay, and the whole will
cave in before long.

Water-shrews are very capricious in their move-ments ; sometimes they follow the edge of the rill, where the vegetation hangs over ; at another time they will come out in the open and swim about. Timid creatures they are : not a movement if you can help it, please, not even a shadow, or they are off to their burrows.

This morning I have been fortunate, for a couple came wriggling out from some cranny, and played about on the surface of the small pool, just below the drain. They appear to dart, but it is really a rapid wriggling motion they make. As they swim they are curiously flattened out. As to their snouts, they are those of moles in miniature, and never still for one moment, but twisting about in every direction. It seems to me, indeed, that what his trunk is to the ele-phant the long snout is to this little animal. When they dive, they look like mice covered with quick-silver. This is caused by the air-bubbles that com-pletely cover the fur when they are in the act of diving ; the fur being compressed close to the body causes the air that is forced out to show like bubbles of silver on the creature. There are two varieties of water-shrews, which differ in size, as do the two water-voles, one being larger than the other. Quiet, out-of-the-way places, where sticklebacks live, suit them ; they are

very particular in their choice in this matter. Min-
nows could not live where they do. A most extraordinary mixture of timidity and ferocity they are.
I tried to pick up one that was making a little over-
land journey, and he jumped, shrieked, and bit in a
most determined manner. Taking its size into consideration, it was the most desperate fighter I have
seen. Battles often are fought between all the creatures here described, in which the weakest get killed.
Anyone wishing to observe them need only go to the
places they are known to frequent. Some who have
read those of our articles in which the exact localities
have been given, have spoken or written quite feelingly because they did not see, in one hasty day's
rush, what it had perhaps taken years to record.
Now and again intermittent calls for 'cold Irish' have
taken up time during that single day's outing, and
drinks taken injudiciously are apt to make the vision
somewhat unobservant.

The little land-shrew, the peak-nosed mouse of
country children, is a very nimble little fellow. He
helps to keep down the numerous insect pests that
would, unless held in check, cause very serious incon-
venience. Much has been written about the mortality
that affects land-shrews at times ; but there is nothing
extraordinary about it. I have known other creatures

thinned down in the same way.  Sometimes I go for
several months, regularly, through the very heart of
the country without seeing a shrew, alive or dead.
The cruel, superstitious folk lore attached to this little
animal which still obtains I cannot deal with in the
limits of this paper.  Prejudice is a stubborn foe
to fight, but superstition is the devil's favourite child,
and he plays strange pranks.  Even now, in this nine-
teenth century, I know of strange rites being practised,
and ones that are by no means harmless.  The
black cat is still in its place in some localities as the
old 'familiar.'  If things are lost in such places folks
do not go to the constable, but to a wise woman, and
dire at times are the results of her insight into utter
darkness.  I have known whole families become
bitter foes through life in consequence of it.  Some
time or other I feel I must write about the folk-lore
and superstitions of the woodlands, for, in spite
of churches and the little Bethels of the elect, not to
mention the advanced schools of the present day,
superstition still holds its own.

Oberon's long-tailed cattle, the white and fawn-
coloured woodmice, with their large dark eyes, are
charming creatures.  It is a pity they do not keep to
the woods; they visit the gardens, and being vegetable feeders with very discriminating palates, like

some other pretty creatures, they have to be dealt summarily with. I feel sorry when I find them in the roads, where they have been flung after having been flattened out by a tile trap. It must be done, though ; choice peas cost money, and the cottage gardener cannot stand his rows being thinned by them.

The harvest mouse, with his grass nest, has been so much written about that he need only be mentioned here.

The field-vole is very like the lemmings in its movements. In Surrey it is called the dog-mouse, to distinguish it from all the others. As large as a half-grown rat it is. I have known this species become a great nuisance in some seasons. They can climb like the others, and in flower gardens they prove very troublesome. The flowers with fleshy stems especially they favour and carry away into their holes. For some time they gave the gardeners much trouble, but the creatures departed in numbers just as suddenly as they came. Whither they betook themselves remained a mystery. A gentleman had asked me to procure a couple of the large species for him : I thought this would be easily done ; but this season (1892) not a single one has been seen about here.

The meadow-vole resembles the field-vole in form, but it is smaller. There is as much difference between the two as there is between a half-grown rat and one that has reached its full size. With the dormouse, or, as our rustics call it, the sleep-mouse, my sketches of small deer must close. Dormice are not nearly so numerous as they were. I have not seen one this year, although I have done a lot of hedge-poking. The town market absorbs them, I fear, for in the country they are now very uncommon. No sleep-mice, in fact, can I hear of, although I have asked about them all round. Folks tell me they see none about where there used to be plenty; and some small birds are disappearing also. I believe many creatures that we can ill spare are sent out of the country, and we shall find out the folly of this presently. The natural course of events has, of course, done much to banish some for ever; the felling of large woods, and draining, for instance. But there ought to be a fine on the shooting of owls; also all the members of the weasel family, as I have said before, should be kept here and not sent to the colonies. Then things might improve somewhat, for, although only small deer, our rodents are capable of much mischief if not kept under.

# THE WITCH OF SMOKY HOLLOW

## CHAPTER I

IN a hollow on the side of one of the hills of a southern county, twenty-five years ago, there nestled a substantial building.  Half farm-house, half cottage it was,
as to style, but a massive structure, having thick oak
doors and window-frames, with stout shutters to guard
the narrow lights.  It gave one the idea that the man
who had built it, whatever he might have been, desired
safety and privacy in his home.  It was grey with age
and covered with lichens when I knew it first ; and
the thinly scattered rustic population in the country
round could give me no information as to who built
it or the character of its inhabitants.

Local tradition, however, hinted that a great man,
who once owned most of the property about, had
caused it to be built for a very beautiful lady to live
in.  At the time I first knew the place it was not inhabited ; and, with the exception of a shepherd or a
woodman, who at rare intervals passed that way, very

few knew that such a house was in existence. The hill-side faced the south, and the only approach was a very narrow path leading from the moor which bordered on the hill-side.

The hollow was wide and deep, and its sides were covered with trees from top to bottom, and tangled thicket growth. A spring of the purest water rose close to the house and trickled along the side of the green stripe which came down from the moor. If you stood on the hill immediately above, you might possibly catch sight of a large stack of chimneys, but nothing more. The manor farm was the nearest dwelling to it, and that was a mile and a half distant.

Though a lonely spot, yet it was warm and sheltered. The venturesome urchins from the hamlet knew that the first primroses and bluebells, to say nothing of the blue eggs of the hedge-sparrow and the larger speckled treasures of the thrush and the blackbird, were to be found in the hollow, and they would sometimes dare to go there in company in the bright sunshine; but they always shunned the spot when the sun was sinking, for they would whisper to one another that 'a summut' in the shape of 'a furrin lady' was to be seen 'walkin' roun' that 'ere old house a-cryin'.'

When wintry winds swept over the hills and

moorlands, driving the snow up in wreaths and deep drifts, then would the shepherd make his way to Smoky Hollow, sure of finding his flock of South Downs there, safe from harm. The spot answered the purpose of a barometer for miles round, for in winter if a change of milder weather was coming you would see the hollow filled with clouds of rising vapour as if the place were on fire. After rain and before rain it rose. That was why the rustics named it Smoky Hollow. In fine settled weather it was perfectly clear and bright there.

''Twud be as well if folks kep' away from that 'ere place,' the people told me : but often I loitered round about it ; the hollow had a great attraction for me. It was, and is still, a birds' paradise, and their song just before spring made way for summer was worth going miles to hear. After a warm shower I liked to watch the various tones of colour the vapour took as the sun flashed through it, while cuckoo ! cuckoo ! cuckoo ! was shouted everywhere, and the scent of hawthorn, primrose, and violet, together with the notes of the other birds in full song, made you feel it was a good thing to live. In all the country round no spot could be found where the nightingale was heard to such perfection as in and around Smoky Hollow.

I confess that sometimes, if I lingered long about the old house, indulging in fancies and speculations as to its inhabitants and the purpose for which it might have been built, a kind of eerie feeling would take possession of me in spite of myself, it was a place so utterly lonely and mysterious-looking.

One day, as I rested on the hill above, a grey old gaffer, bent nearly double, chanced to come in sight. He lived at a cottage situated in a dip of the moor. I thought he ought to know something about the house. His poor dim eyes winked and blinked with pleasant expectancy when I asked first if he smoked.

' Sure I does, when I kin git a bit,' he answered ; ' but 'tain't often, and when I does it ain't up to nothin' like.'

His hands, knotted and wrinkled with the toil of more than seventy summers and winters, trembled with eagerness when I asked him to share with me the contents of a pouch filled with genuine old Virginia. Sniffing at it several times, he said :

' Ye wun't mind me hevin' a whiff now, will ye ? I can't keep from this 'ere bit o' 'baccer nohow,' at the same time producing a black pipe about an inch long from some portion of his garments where it was coddled up. Next he fished up a regular tortoise-shaped steel tobacco-box, which he filled carefully, anxious not to lose one shred. Then he filled his precious nose-warmer of a pipe, very lovingly patted it on the top, and began to fumble for a lucifer.

'Thank 'ee, 'tis real kind on ye,' he said, as I held him one ready lighted ; and, dropping his stick, he seated himself on the bank beside me, for the day was warm and dry, and began whiffing away in the peculiar manner of a hardened old smoker ; not hastily, but making the very most of every whiff. It would have been positive cruelty to have disturbed the feeling of supreme content which his countenance expressed. Presently he informed me that he 'ain't had a bit o' baccer like this 'ere, never in his martal life afore.' When I told him I had been down in the hollow by the old house, he said he had heard 'as how the inside o' that 'ere place was kep zackly as 'twas when the furrin lady died. There be martal strange tales 'bout that place,' he added. By degrees I got from his lips the following story, which I found afterwards was correct in the main points. I give it in his own dialect.

'The squire as owned the land about here was 'bliged to marry somebody as he didn't care a rap fur. 'Twas a matter o' jining two properties, so far as I could mek out. He was a fine feller, so I've heard, an' the lady as he did marry raly was fond o'

him, though he cared nought fur she.   He'd bin
travellin' in furrin parts, an' fell in love like wi' a lady,
and wud ha' married she, but the old squire, his
father, swore a most desprit wicked oath that if he
did it he would cut him off with a shillin' to buy a
rope fur to hang himself with ; he should niver darken
his doors agin as long as ever he lived.   Well, he wus
'bliged ter du it, an' he married the lady 'as his father
wanted him tu.   But he waunt happy ; no more wus
she when she found as she'd no chance o' makin' him
fond o' her.   An' what duse she du, mad with temper
like, but tell him 'twas a real pity his lady love over
the sea waunt thear tu see how happy he wus.   'Tis
said he said summut back to she as made her shrike
out like a mad critter, an' when her maid run tu her
she found her in a dead faint, an' her husband gone.

'Where he went tu waunt known about here.
Arter a time he came back, an' 'twas whispered very
quiet like, as he had somebody cum with him.   His
wife had left these parts then, an' gone tu her gran'
house in Lunnon ; but I hev heerd tell as she'd walk
about her gran' rooms many a time cryin' most pitta-
ble.   They niver had no children, but the lady as
lived here had.   Arter a time she tuk ill an' died.
He carried on dreadful then, 'twas feared fur a time
as he wud lose his senses.   Arter the wust part o'

the blow had left smartin', he used come an' wander roun' that 'ere house in the holler fur days, an' nights too sometimes, an' then he said as he'd travel, there waunt nothin' tu hold him here now; an' he give the most strict orders as no one wus tu live thear, only an old critter tu see as nothin' went wrong inside like, an' the room as she wus fond o' sittin' in wus tu be locked up, an' iverythin' wus to be left as she'd left it; nobody wus tu lay a finger on it. He took the children with him when he went to furrin parts. He niver cum back, fur he died thear, an' then his wife she fretted an' troubled so, that 'twaunt long afore she went off; fur she wus martal fond o' he.

' 'Twus said not so werry long ago as some o' that 'ere furrin lady's family wus livin'; an' some of the folks 'bout here, older than I be, say as they reckons as they old chimbleys ull hev smoke cum out of 'em afore long. An' that is all I knows on, fur tu hev any sense like.'

.    .    .    .    .    . .

## CHAPTER II

In the spring of the following year my business took me again into the neighbourhood of Smoky Hollow, to stay for some time. The birds sang their loudest as they made their nests, and the boldest of the children again ventured within the hollow; but, to their terror, two of them, on getting there before the others, caught sight of a woman standing close to the house, the door of which stood wide open. Making a sign of warning to their companions, they crept through the thicket and wood again, on to the moor, where, with one look at each other, they scampered off like hunted rabbits, their hearts in their mouths, but no word on their lips. These forest children are peculiar in one thing, they are more silent when on the hunt. Each one made tracks for his own home without speaking. Once safely there, the tongues began to wag again, and they told their parents that 'niver agin wud they go to that 'ere place, fur 'twus hanted; 'twus right, ivery word on it what folks said, fur the ghost o' that ere furrin woman, as folks talken on, stood thear now.'

They were cross-questioned, but all told the same tale. The father of the biggest urchin was hard of belief, and, much to the boy's disgust, he rubbed a dose of ash-plant oil into his back and shoulders, telling him between each application that he'd 'larn him better than tu give his mind to lyin'.'

'Go, an' see fur yurself!' yelled young Hopeful, 'an leave off quiltin' me.'

His persistency impressed the father, so that he made his way to the spot on the crest of the hill overlooking the hollow. There he saw the smoke unmistakably rising from the great stack of chimneys. Muttering to himself, 'Where there's smoke there's fire, an' where there's fire there's some one to light it,' he started back again quicker than he had come.

The same evening he was the hero at the public-house, where he repeated the story, saying over and over again, 'The thing hed cum tu pass, as I allus sed they wud, whoever lived long enuf tu see it. Wut is tu be wull be, in coorse o' natur, an' nuthin' ken stan' agin it, nor yet perwent it ; fur mark my wurds, if summut hain't cum back agin, an' I've sin it.'

For the latter fact he drew on his imagination, as he forgot he was indebted to his son for the information concerning the lady. The company were edified, and asked him to 'hev a wet,' over and over again.

A few days later his story was confirmed. A bad accident happened to one of a party of woodmen who were felling trees in the neighbourhood of the old house. A mate's axe, glancing off from the trunk of a tree, struck him on the arm, cutting it badly. The pain and loss of blood being great, they

M

made a rough litter and hurried home with him, after binding up the wound to the best of their ability. As they passed the entrance to the hollow with their burden, they saw a strange woman standing near the house. She came to meet them, and in excellent English, but with a slightly foreign accent, questioned them about the accident, and bade them rest, as their injured mate had fainted. Going indoors, she returned in the space of a few minutes with bandages, and followed by another woman, who looked like a servant, carrying restoratives.

The wound was quickly and skilfully dressed; then she made the poor fellow swallow something which she gave him, in a small phial, and told his companions they could now carry him further without undue haste, for he would not take any harm. Before the astonished woodmen could thank her for her kindness she had vanished. As soon as the man reached home the village doctor was sent for. When he saw the bandages he asked a few questions; did not remove them at all, but told the man that the lady who had put them on must have had great experience in such matters. Then he cross-questioned the other men as to the strange lady's appearance. The description they gave was, 'not a young un, tall an' handsome, with dark hair an' eyes, leastways her hair was grey, a bit ; an' she'd a nice voice an' way o' speakin'.'

Whatever the doctor thought, he said nothing.   In the evening when they went to the public, as they generally did after any unusual occurrence, they found the hero of the night before giving a most minute description of the lady's dress and features, over copious drinks.   After listening to this they told him that he was ' a paltry lyin' varmint, and if he waunt out of the place in a jiff, he should make acquaintance with some shoe-leather.'   Sneaking home, where he found his son had not yet gone to bed, he gave the poor lad another dose of ash-plant oil, telling him ' the next mare's nest as he found he'd best keep his mouth shet tight about it, an' not git grown-up people inter trouble.'

By degrees the simple country folks lost the vague feeling of fear they had at first, and if they met the lady near or about her home, would lift their hands to their heads in token of respect ; for many an act of kindness did she show them in the most quiet manner.

At last the nearest resident clergyman paid her a visit ; he was courteously received, and every mark of respect shown him, but his visit was never repeated. A small farm near supplied almost all the house

M 2

required, and the farmer's only daughter—fair-haired, blue-eyed Annie—carried the produce to Smoky Hollow. The lonely woman always gave her a warm welcome, and soon won Annie's confidence, so that she told the lady all her little private affairs, and used to look forward to the day when she paid her weekly visit, the refined and gentle conversation of the stranger being a great treat to the country girl.

A poor young girl, who had been what is termed unfortunate, whose little one was ill, and had been given up by the doctor, went one day to the farm to ask if Annie would get the lady of the hollow to see her babe, as she had heard she was good and wise and could cure complaints. Annie bade her go to her at once herself, and take the child ; she was so kind she would turn no one away.

On the way there the girl met her and told her of her trouble. Lifting the light covering off the child's face, the lady looked intently at it for some time, and then said :

' He will not live. It is better so, for he will go to our Father. Before the week is over he will be with Him who said, " Suffer the little children to come unto me." '

Crying bitterly, the poor mother, who was only a girl in years, turned to go ; but the lady stopped her, and placed something in her hand, saying, 'The dying and the dead are beyond our aid, but it is our duty to comfort and assist those who live and suffer ; is it not written, Much is forgiven her, for she loved much ?'

The child died, as she said it would ; and the old crones about said the lady must be a witch, 'but a rale good-hearted one'; for she had given the poor girl far more than enough to pay for the funeral. The babe's little blue coffin was ornamented too ; those who live in country villages know how dearly they love to pay this last mark of respect to the dead, and to be able to do without the aid of the parish authorities in that matter.

After that the lady was named the Witch of Smoky Hollow. Her real name was never known by the rustics.

Near the family vault of the old squire there was a flat slab of red granite ; on it no name or date was inscribed, only a simple cross had been cut in the hard stone.  One Sunday the villagers gathered round it with wondering eyes ; for a large and beautiful cross of white marble had been placed at the head of the granite slab.  Passion flowers and leaves, most beautifully chiselled, adorned it, and at the foot of the cross were the words, 'Jesu, mercy.'  Who had placed it there was not known.

Annie presently found a true friend and comforter

in the mysterious lady, to whom she had already told her simple love-story. William, the lover, was a smart-looking young fellow, second gamekeeper to the squire at the Hall. Latterly Annie had looked pale and lost her brightness, and her friend, noticing it, soon drew from her the cause of this. There had been much company at the Hall, and some of the 'ladies' ladies' were very good-looking and smart young women, and William, who had often to go up to the Hall, was noticed by them as being a very well-built, fine young fellow. So they made much of him when they had the opportunity, and poor William rather lost his head over it; so that if things did not alter there was great danger of Annie being made very miserable, for she and William were really engaged to be married. She told her story at Smoky Hollow with a tearful face, and asked the lady what she ought to do.

'Cheer up, little one,' said her friend; 'if he is worth having, he will come back to you gladly before long. I have heard they call me the Witch of Smoky Hollow. Well, I will give you a charm that will bring him to you again. Come, let me see you smile a little, then I will go and prepare for you my charm.'

The charm was a sealed packet, which she bade Annie place under her pillow, and not open until the

morning.   The girl religiously observed these instructions.   In the morning she unfolded the paper and
found only these words written :—

> ' A fickle lover will prove a faithless husband.'
> ' Lightly won is little valued.'
> ' That which is always to be seen is little looked for.'

Annie had enough woman's wit to understand the
meaning her friend sought to convey to her ; and she
asked her father to let her go and pay a visit to some
relatives at a little distance.

' Ay, do, lass ; fur mother an' I ha' noticed you
bain't so peart as you used to be.   Have a month on
it ; ye niver mind fur money, have whativer ye like.
'Twill all be yours some day, ye knows.'

Next morning Annie was gone.

Soon William found that the ladies' ladies grew
tired of his company ; he had not the refined and
aristocratic airs of those superb creatures Jeames and
Henery from Berkeley Square, and, as one of his
admirers confided to her friend, he was ' gawky, in
fact, my dear, he is nervous in our society ' ; and it
then dawned on the young man's slow perceptions
that he had been a fool.   His conscience smote him
too, for he was good at heart.   Dejectedly he made
his way to the farm and asked for Annie, but learned
to his dismay that she had gone away.

'We ain't sin much of ye lately, William,' said her father; 'no more ain't Annie. Have ye bin night-watchin', or what?'

Weeks passed and no tidings of Annie reached the penitent lover. Then a note came to her in the same handwriting as that on her 'charm,' stating that William had given notice to leave his situation, and intended going as a soldier.

Before his time of service expired, however, going on his rounds one day near the hollow, he met Annie. That evening he returned to the farm with her, and the kind farmer asked for no explanations, but greeted him with a pleasant speech of welcome:

'I be glad to see ye, son William, that I be,' and the mother echoed his words.

On their wedding-day, when they returned to the farm from church, Annie was told that a lady wished to see her alone for a few minutes. To her surprise there stood the lady of the hollow dressed in the black garb of a Sister of Mercy, and from her neck, suspended by a golden chain, hung a large crucifix. Taking the blushing girl by the hand, she said, 'We may not meet again on earth, little one, for I have not long to tarry here. Accept this packet for a wedding present; and this coral and amber necklace place round the neck of your first-born.

The clasp of it is said to be a charm against witches ; and when you and your husband see your little ones growing up round you, remember the Witch of Smoky Hollow.'

Before Annie could compose her feelings enough to thank her, the lady was gone.

.     .     .     .     .     .

That same day the vicar of the parish was summoned to go to Smoky Hollow. What passed there was never known, but from that day the poor received benefits twice a year regularly that they had not had before ; and to this day the beautiful marble cross is religiously cared for, and the poor folks round bless the memory of the Witch of Smoky Hollow.

The lady was never seen again ; she departed as mysteriously as she came.

.     .     .     .     .     .

One Sunday morning the congregation at the parish church noticed that their vicar was deeply moved, as he spoke of 'one who had gone before.' A letter had reached him with a foreign post-mark on its black-edged envelope. It came from the Lady Superior of a convent in Italy, to tell him that 'Sister Mary' was dead.

## *LITTLE JAKE*

ON a glorious afternoon in September I stood on the old Roman camp-ground, looking on a scene not to be surpassed in England. With the exception of the crests of two near hills which break the line of sight, the view is almost uninterrupted, and extends over the weald of Surrey; Sussex, with parts of Kent and Hampshire, are in the distance, lost in a blue haze.

Hills and valleys, woods and waters, cornfields and farm-houses, with slender church spires here and there pointing heavenward, satisfy the eye. In the foreground the purple heather in full bloom is mingled with the golden flowers of the furze. The whortle-berry bushes are getting orange and crimson tints mixed with their myrtle-green leaves ; the dewberry and blackberry sprays are gorgeous in colouring of gold, crimson, pale grey green and warm olive ; and the whole is framed in a setting of dark encircling firs. A few strips of land quite near are bare ; .

there the great forest fire left its mark, but even here
the grass is springing up fresh and green, and the
fern fronds are showing their curled tips ; vegetation
springs up like magic from the warm earth moistened
by a few showers. The heather makes a good couch,
and, as I lie resting here for a while, the sound of a
sweet chime of bells comes up from far below me.
The day was drawing to a close when I reached the
churchyard in a valley. I was looking for a little
grave, that of a child who had been a favourite with
me in the woodlands some years ago.

'That is the one, sir, with the flowers all over it,' a
little girl replied to my question. She was passing
through on her way home from school. Looking
shyly up at me, she added, 'Did you know him? he
used to come to our school.'

'Yes, my dear, I did know him, and I have come
from a distance to look at his grave.'

She surveyed me in an artless, pretty fashion, as
only a child can ; then, bidding me good-bye, passed
on, leaving me to muse beside the flowers. The
place was quiet and full of repose, as befitted a last
resting-place ; and I stood there for some time
dreaming of the past, and living over again scenes
which had left good pleasant memories behind.

'Ay, 'tis the best we can do for him,' says a voice

close to me.   I had not heard any footsteps over the soft green turf, and I started, turned, and saw the father of my friend little Jake close to me.   Shaking him by the hand, I tell him it is the best, and he has done it well.   This little lad was barely ten years old, the flower of the home.

'Little Lucy, our neighbour's child, told us some one had come to look at Jake's grave, and from what she said about his having come a long distance I made sure 'twas you.   Times and again I've wished I could see ye once more.   I'm glad you've come.'

'Would you mind telling me about him ?   Or would it hurt you to talk about him ?   If it would, let it be just now.'

'I can speak of him now without smartin', the time has bided by for that.'

'Come up a piece of the valley with me, then, and tell me what you can.'

'Well, ye see, he just faded away like—faded clean away ; there ain't no other words for it.   He used to come reg'lar to meet me comin' home when work was done.   Pleased as possible he'd be to carry the basket or something ; and he'd chirrup away about what he'd bin doin' at school, and other little bits o' things, an' he'd ask me where I'd bin, and what jobs I'd bin at.   One night I missed him, an' when I got

indoors he was there sittin' in a chair.   Mother said
he was poorly like, an' he was very quiet : we
missed his little chatter.   The next day he was no
better, an' then we had the doctor.   Things went on
for a week or two the same way, sometimes a little
better, sometimes worse.   When I asked him how he
felt, he used to look up at me an' say, " Tired, dad, so
tired ! "   I tell ye somethin' used rise up in my throat
when he'd speak like that, all chokey like.   He'd talk
to his bullfinch a good deal ; an' the creetur knowed
him like a human.   It was from you he had it, you'll
remember.   Ay, he did think a lot o' that bird, an' the
creetur was fond of him as possible.   'Twas a cruel sight
to me for to see his poor thin fingers play with it ; for
he'd begun to waste.   The doctor was good an' kind to
him ; uncommon good he was, an' he'd chat to him
cheery like whenever he come.   One day I see him
get up from where he'd bin talkin', an' walk away to
the winder ; an' he bides there a bit.   It warn't a
fly as he wiped off his face there.   May he be
rewarded for all his kindness, in a different place to
this !

   ' At last he took to his bed reg'lar ; he was so
weak.   We had to take the bird up where he could
see Jake, for it pined an' mourned so when it missed
him.   Directly we took him up an' he caught sight

of Jake, he trimmed his feathers an' was peart as
possible, and never pined a bit. It took the heart
clean out of me, for I could see he'd be goin' before
long. I used to go to work an' come home agin like
as if I was in a dream. The end on it come quicker
than we reckoned for. One evenin', a bit before the
sun went down, we was up in the room, mother an
me, talkin' to him. 'Twas bright as gold, the sun
was, an' the tops of the firs was peaked out clear agin
the sky ; when he says, weak, an' low like, " Dad, lift
me up, an' let me look out of the winder."

'I lifted him up, an' took him to the winder. He
laid there for a time—ah ! so quiet—lookin' out.
Then he looks up in my face, an' he says, quite peart
an' lively like : " Dad ! it's gettin' lighter, and—*I*—
*see*——" They were the last words of my poor little
Jake, for he died with them on his lips : yes, he lay
dead in my arms. Whatever he had a sight on, it
made him happy ; he had a smile on his face, an' he
kept the same look on him when he laid in his coffin.

'Mother thought for the minnit he'd gone to
sleep ; he went so gentle like. It broke us both clean
down for a time—right clean down, it did ; but that
has passed, and the smart is gone, and we hope for
the best, though 'twas hard to think so then. We
did all we knowed for his bird for his sake, 'cause

Jake was so fond of it ; but it never cheered up, it missed him so.

'One night I come home feelin' dreary like, and there was a hankercher over the cage. Mother was cryin'! It don't take much at such times to upset ye. I went and took it off, and what do ye think I see? The bird was dead, with his breast agin the bars of the cage, just in the same way as when he used to watch for Jake.

''Twas only a bird, I know, but it did hurt me at the time. . . . Things comes clear to ye at times. When you first bided about here, before we knowed anything about ye, times an' agin have we sin ye lookin' at the sun settin' over they firs. Since Jake's bin gone, mother an' me have thought on it, an' spoke of it to one another, that perhaps some one as you once knowed faded away westward, as the sun went down.'

## *IN FLIGHT TIME*

SUMMER has gone so very gently that the only
visible signs that the time of the falling leaf has come
are the changes in the foliage of the woodlands, the
gathering in of the crops, and the flight time of the
birds.

A rolling sea of mist ascends from the valleys,
creeping up the sides of the hills to melt away as it
reaches the uplands, where the sun is shining brightly
above. Those who have not made the flight of birds
a particular study might fancy these wood-covered
hills, great stretches of uplands covered with the finest
turf and sandy heaths, would not be the localities
about which to watch for them at this time; yet
these are the very parts they pass over, and where
some of them rest for a time, year after year. Divers,
swimmers, and waders pass over from the north-east
to the south-west and south, in the morning, at mid-
day, and in the night-time. The birds that break
their journey find in this line of country all they

require—rivers, brooks, swamps, pools, and ponds, mile upon mile of broken ground and rough cover and, above all, solitude.

Come with me to the woods that cover the tops of the hills as far as the eye can reach. Great open downs break the masses of woodlands; but look where you will, around or in front, in the valleys below or in the far distance, there are the woods which, in past times, reached, almost unbroken, from London to the south coast. Wanderers from the beaten tracks who are familiar with this country can travel over old bridle paths, still open, leading from one county to another. Breaks occur here and there, but they can be picked up again; and only in this way can the real beauty of the district be rightly seen and understood.

Some of my critics have observed that if the exact names of localities referred to had been given, our readers would have been better pleased. So, for the benefit of such, let me map out one day's good walk, supposing the would-be naturalist to take the first train down from London, which reaches Dorking at seven in the morning. From the railway station a path leads up the hill to Ranmore Common, over which a good road runs to the edge of the wooded downs which stretch as far as Guildford; and if the

N

track through the woods be followed, Newlands
Corner will be reached, from which point a view will
be seen that I could not here do justice to. From
this a path leads down to Shere, to beautiful Aldbury
and Chilworth.

When the line is crossed at Chilworth station, a
green lane leads to Black Heath and Farleigh Heath.
Then, if my reader cares to travel by a track through
the wild land, which runs about the centre of it, he
will have Ewhurst Hill, Holmbury, and Leith Hill on
the right, St. Martha's chapel-crowned hill and the
high downs leading to Box Hill on the left ; and so will
arrive at Dorking station again, after a day amongst
one of the many beautiful hunting grounds that the
naturalist may find amongst our Surrey hills. But,
unless he be a good walker let him not attempt such
an excursion.

Nearly all the birds that live in and about the
woods and uplands have now—late in September—
completed their moulting, and their young ones, in
their first plumage, accompany their parents.

By many species the large stubble fields, with
their high old hedges, which border the uplands, are
frequented. All wild fruits and berries are ripe now ;
in fact, that universal favourite, the mountain ash, will
be found completely rifled of its crimson clusters and

having only the stems left ; even the berries that have fallen are eagerly sought for. Great emmet heaps are numerous round the dry, sloping edges of the fields ; so are old blackthorns. The yaffle, or green woodpecker, is by no means scarce here ; you may put up one or two families of them as you walk along—old birds and their young. These latter are so much speckled that, with only a casual glimpse, they might be mistaken for young missel-thrushes. Only for a moment, though, for the yells of the parent birds and the half-choked cackle of the others speak for themselves. If you could take one in your hand to examine it, you would find it in rare good condition ; grubs, caterpillars, beetles, and ant-eggs, with other matters, have made all the family right plump.

Hawfinches are in full force just now ; they do not harass that afore-mentioned old friend of mine, grinding up his precious marrow-fats, since the season of these is long past. They are prospecting about with their keen eyes to find out which crop of berries or fruit will be ready for them first.

Under the wild cherry trees they will search for the stones of fallen fruit ; then the damson trees will be visited in every cottage garden. So they jerk from one place to another, continually on the lookout for something wherewith to please and satisfy palate and stomach.    The birds of the year resemble small parrots, for their plumage, which is very like that of a dull-coloured greenfinch, is now broken into by some of the markings of the old birds.    As the winter draws near, they will form small families of ten or a dozen, as the case may be, and three or four families in a bunch, to range for miles over all the rough as well as the cultivated parts of the land, returning to their resting-places in the evening.    To a certain extent this bird is local ; this being influenced, no doubt, by the food supply.    Wild fruits and berries are their portion.

The corn-bunting I have only met with twice in this district during a period of twenty-five years.    I put it up then from some rough 'torey' grass which was dotted over with thorn bushes.

Linnets frequent the fields in flocks, so do yellow-hammers.    If you stand perfectly still you can watch them feed.    A great quantity of grain, both wheat and oats, is scattered about over the ground after the crops have been carried ; but these grains, while they last, only form a portion of their daily food.    The seeds of weeds, which are the pests of the farmer, are what they subsist on mainly ; which diet is varied by certain insects, on which they prey in all stages of their life.    It is a fact well known to naturalists that all birds, from the raptores down to the swimmers and waders, feed more or less on insects, according to the seasons. Some of the larger beetles are eagerly sought for when flying or resting, and, of course, during the nesting season the young of the greater portion of our British birds are fed exclusively on insects. Look only in the mouth of any common bird when it has young to feed—some rook, starling, blackbird, thrush, or sparrow, that may have been wantonly shot—and you will never allow another to be shot when nesting, if you can prevent it. Of course, as I have said before, the birds will, when their time comes, take their share of the garden fruits and vegetables; that will, however, be when they have done nesting.

But birds are now flighting to and fro in numbers; some of them migrating in earnest, others indulging in those restless movements which are peculiar to some before they take their final flight. The stone-curlew, the great plover or thick-knee, is one of these. He frequents the great flint-cumbered fallow-fields bordering on the uplands. From these he flits, as the dusk comes on, to the close-cropped sheep-walks on the downs, where he startles the shepherd's boy with his mournful cry; the note is peculiar, so is the bird's flight.

' Have thee heard the stone-curlew lately, Tom ? ' I
ask one of my friends.

' Do ye mean that queer bird what hollers out so
o' nights, an' skims along the ship-walk when the
dims come on ?   I've heerd un, and sin both on 'em
'long with the young uns lots o' times since I sin ye
up here.   Why, they hed their nest out in that 'ere
flinty faller-field.'

' I told you they were nesting there.'

' How did you know ?  You ain't up here same as I
be.   They hed young uns in that field ; old Jack, our
ship-dog, picked one on 'em up.  Two on 'em with the
old uns there was ; the bird warn't hurt, fur old Jack's
middlin' tender-mouthed when he's on ship-ground.
He's bin larrupped when he wus young for breakin'
the skin o' the ship, an', bless ye, he don't do it now,
cept rabbuts an' varmin.  He grips them right through,
I can tell ye.  So thet outlandish-lookin' critter didn't
hev a feather pulled out on it.   It hed eyes like them
yeller-eyed owls.   I put un down arter lookin' at it,
an' it could run most 'mazin'.  If I'd ha' known where
to find ye, I'd ha' kep it for ye.   But 'twas best to
let it go ; them 'ere queer old birds would holler at me
o' nights when I come over the ship-walk if I didn't.
They sounds wheesht like o' nights.'

I tell him they will soon be gone, that their restless flappings and wild cries tell of their early flight to other lands. A few remain in England, but the great body of stone-curlews leaves us.

From Tom's 'ship-walk' there is one of the fairest sights in England. Great stripes of velvety green turf from half a mile to a mile in length, and of widths varying from fifty yards to two or three hundred, all dotted with sheep. From these stripes you dip down into green hollows, sheltered from all the winds that blow, and studded over with fine clumps of juniper trees, heather, and furze in full bloom even, yet, though it is late autumn. Where the ground, has been broken in past times for stone-getting, the stone-bramble, as some call it, or dewberry droops down with its large refreshing fruit. I have seen some of them in pits and hollows that were quite as large as ordinary mulberries. Mushrooms stud the turf of the sheep-walk, but as our business is with the birds only, we leave these for Tom to pick up. The badger, too, roams along the sheep-walks in quest of mushrooms, slugs, snails, and other small matters. Here one might wander for weeks, and yet find fresh beauties every day.

So still is it to-day that the thistle-down hardly floats clear of the dog that stands near the plant it fell from, and old Jack actually pants with the heat.

At long intervals a shivering sigh passes over the woodlands, dying away as quickly as it came. The first sigh of a declining year is this ; only at one season is that mysterious sound heard, and it is that of the falling leaf. Some may not have noticed this, but to many of us it is one of the unmistakable warnings that summer has flown, and that the beginning of the year's end is near.

But our musings are put an end to by Tom, who shouts, ' Massy alive ! what be these 'ere things a-comin' Mister ? Look at 'em with that 'ere tool o' .yourn ! '

I look in the direction indicated, and see nine large birds coming at top speed, their wings full spread—their best migrating rate this is. They are herring gulls, all of them pure grey and white. As I hand Tom the glass they are right overhead. After a moment's silence he yells out, ' As big as geese ! I ken see the feathers in their wings, an' their beaks an' eyes ! Why ye ken a'most touch things with this ! '

Those who know the ways of the herring gull know well that they visit the cornfields at times. Although this place is full thirty miles from the tide ' as the crow flies,' and the gulls were making direct for the south coast, it is possible they may have rested on some of the large sheets of water that so

plentifully dot this line of country. Round the edges
of the lonely pools in upland hollows the print-marks
of webbed feet have certainly been found, and now
and again the birds that left these have been fired at.
From a minute description given me of some such by
one man who fired but missed, one who was not well
up in bird life, I believe them to have been young
birds of the lesser black-backed gull species.

Wild geese, to my certain knowledge, visit the
stubble fields after the corn is cut; so do the ducks.
I am quite inclined to believe, from what I have
heard and seen, that gulls break their flight to rest
in some lonely swamp, or on some sheltered woodland pool.

Flight time gives one much to ponder over. The
birds come and go, but there are no set rules for
them. To a certain extent they must be the creatures of circumstances over which they have little
control. Food and shelter are needful to the well-
being of all wild creatures, whether they be high or
low in the scale.

Some who rarely move out without a gun in their
hands, mourn over the decreasing numbers of wild
things. To these I would say, 'Put the gun in its
case, and take out a good field-glass in place of it;
there is far more pleasure in watching the life *in* a

creature than in knocking the life out of it. The gun
may, however, in some cases be a necessity; that is,
of course, a different matter. But I can honestly say
that any real insight I may have into Nature and her
ways began when I made that exchange myself.

Herons are moving now, young birds chiefly; any
pond or pool that is on waste ground, and the
rougher the better, will be visited by them. Most of
these ponds have fish, small carp as a rule. How they
got there is a puzzle which I will try later on to make
clear. One heron, passing over some fields, was
mobbed by a lot of rooks that were about to settle
there. They went for him with one accord, darting,
cuffing, and striking in all directions, and cawing and
quarking at a fearful rate. The heron wobbled and
tumbled at first like a lump of feathers gone wild.
Presently he cleared himself; then he circled in the
air, the rooks following. Again and again he circled,
higher and higher above his persecutors, and these
tried to reach him, cawing their loudest. When the
heron had mounted so high that he looked the size of
a rook, he became stationary, and flapped his wings
like a large hawk. Then he changed his tactics, and
there was a transformation. The rooks became
aware of it, and with loud cries of alarm they dived
and tumbled in all directions, darting down low over

the ground, and then up into the trees like rockets ; there they hid themselves in perfect silence. I watched the incident through my glasses from the start to the finish, and was able to see some fine flights on the part of the heron, as well as by the rooks.

The latest detachments of swallows are preparing for flight. For six successive mornings I have been taking note of them near one of their favourite gathering places, which is visited by the sunlight directly the sun rises over a neighbouring hill. An interesting sight it is as they sit in long lines, tier after tier, on the postal-telegraph wires. With very few exceptions these are all young birds ; they do not move as we walk gently beneath them, only twitter as they preen their feathers.

All are swallows, and yet no two of them are exactly alike. The parent birds dash down to feed some of the smallest, without, however, settling. These companies are composed of first and second broods ; the weaklings will be the latest to leave—that is, if they live. A sudden change of temperature, such as we so frequently have, from heat to frost, kills them in numbers. Some half-dozen old birds we note, distributed at long intervals among several hundreds of birds, apparently just to maintain order ; for I notice that when one of the stronger young fellows flies from his perch and tries to hustle a weaker one from its place, he is quickly brought to book by one of· those older ones, so that, on the whole, it is a very orderly gathering.    The coming and the going of swallows is certainly a very remarkable matter for consideration, for they come and go by twos and threes, in small companies and in large ones, to fill their accustomed haunts by degrees ; quitting them in the same gradual fashion.

My daily avocation compelling me to be about all over the country in every sort of weather, foul as well as fair, I see many an incident which might otherwise have escaped observation.    In the early part of April, this present year, 1892, business took me through some fine park-lands just then lit up by the sun.    When I reached the centre of these a mob of swallows, some fifty or sixty of them there were, dropped down from their migrating flight to settle on the park railings.    They were in splendid plumage, and apparently right glad that their journey was over, for they twittered, broke out into full song, caressed each other, and preened their feathers, fluttering over across the road and back to the rails again.    So tired were the beautiful, innocent creatures that they allowed me to walk along the whole length of their line, and to look at them at a distance of barely six feet from this their first resting-place. When birds that are on migrating flight settle, there is no preliminary inspection of localities or surroundings. They simply drop down evidently dead beat, just as we should, under similar circumstances, rest on the first seat that presented itself. On the following day the swallows were to be seen in their accustomed haunts, darting over the village green, shooting under the grey arches of the bridge which spans the Mole, above the water-meadows, and again circling round the elms that surround the old farm.

## LONGSHORE MEMORIES

NEARLY all the fauna of England were represented fully in our fens or marshes—the red deer and the roe deer excepted. Foxes, badgers, otters, martens, polecats, stoats, weasels, rats, mice, voles, shrews, and moles were all gathered together in one district. Fish, of course, and reptiles and insects were there also, in abundance. In all the fens or marshlands where the water is free from salt, rank lush-vegetation will be seen in its perfection. About those spots where the salt water mingles with the fresh at times, owing to high tides and the natural creeks that run up to meet the inland streams, the tangle will not be quite so rank. You may get through it, with considerable pains, in certain places. No wonder that such localities, before they were drained, should have been a refuge for all wild creatures.

Here the birds of prey literally held high revels; and, in what was their paradise, the bittern boomed or bumped, the herons quarked harshly, and rails, coots, grebes, and snipes sent forth their voices to mingle with the croaking of the frog. To many of us this was the music of the marshes and a delight to the ear. Hundreds of birds, rare ones many of them would be considered now, though common enough then, were shot without exciting the least notice or comment—bitterns, herons, avocets, ruffs, and reeves, as well as all the waders, swimmers, and divers. As to the peculiar sounds heard at certain seasons, it would be difficult to give any adequate idea of these. It was only natural that men who were shaken by ague, and half delirious with the marsh-fever, that comes in winter time as well as in summer, fancied that supernatural creatures were abroad. One saw the wrecks of what once were strong men and fine healthy women, prematurely aged by ague, fever, and the large quantities of laudanum and brandy by which they sought to keep the former foes in check.

Although countless larks soared and sang in the bright sunlight, strange unearthly cries sounded through the long nights. Many places that had evil reputations were literally death-traps to the man who placed his foot on one of them, fair though they were for the eye to rest on.

The only literature to be met with in our smaller

dwellings, if the owners of these could read at all, would be a large-lettered Bible and an almanac of the plainest form. If this predicted future events, or pretended to do it, so much the more was it appreciated. In lonely homes, often cut off for weeks and months together from their nearest neighbours, when the waters were out and the land was drowned, their position was at times an unenviable one, and under the most favourable circumstances their lot, in the light of the present day, would appear unendurable to many. Yet these served God and did their duty, deriving as much, and perhaps more satisfaction therefrom than do the more highly-favoured ones of their class to-day.

The simple arts of reading and writing were highly valued ; those mothers and fathers who could read and write taught their own children and those of their less gifted neighbours during the long winter evenings. A man or woman who could read a six-weeks-old newspaper to the neighbours, and write a letter for them at intervals, few and far between, was much looked up to and respected. A book, especially if it was illustrated, would be lent for miles round ; and when, through constant wear and tear, it began to go, it was stitched and pasted in wonderful fashion.

Little in the way of fine art reached us, beyond the gay pictures on the top of the pretty fruit and glove boxes brought for their mothers or sweethearts by those young sailors who had visited France or Holland. But though our folks had small book knowledge, they learned much from nature direct. The four seasons brought to them little variation ; one year was to them the same as another. In spring the waders wakened up the flats to life again, when they nested in their wonted grounds. The pewits ran about, taking little heed of man, woman, or child. One of their favourite breeding stations, I remember, was close to the most frequented track of the flats.

The boys would give you the action of the snipe in breeding-time, as he mounts up piping, or, as they termed it, 'whinnying,' as well as the humming sound he makes in his descent. The action of the bird's wings they would make with the hands, as they imitated the bleat and the hum to the life. All the various cries and motions of the wildfowl they were familiar with ; from the quick, rocket-like spring of the teal to the heavy flap of the wild swan rising from the water, from the little grebe to the great sprat diver, they knew them all.

The number of churches along the shore was remarkable, considering the thinly scattered population,

O

but I fear the number of the orthodox faithful was very small. My own people were attached to the ' Hew Agag in pieces ' school, and, as a rule, I was taken to hear their favourite preachers. I remember well the fervour with which these depicted the horrors of the infernal regions, to the edification of such as had scapegoats, as they always called the scapegraces of their families—possibly with more truth than they imagined. As a boy, I have shivered with fear and perspired in real agony under some of these discourses. And yet there was another side also to this, and in the annals of our fishing village I have told how the influence of some of those unconventional Christians affected us at times in a more healthy manner. On Sunday you would see a few figures moving across the marshes, patiently trudging for miles on their way to church or chapel ; grave men and sedate women, and the plovers crying and flapping, unheeded, about them as they walked.

The beautiful grey and white gulls, resting in and around the clear shallow splashes, barely drew their attention from the path they were travelling by. These were so mixed up in their daily lives that they would only have noticed their absence. When the gulls rested like that, and the pewits ran and flapped all around, they knew it would be settled

weather for a time. If they hovered and cried overhead, and made for the upland fields, foul weather
was certainly coming, and they must prepare for it.
When some strange sea-fowl, one that was an unusual
visitant to our shores, was found in the fields, there
would be much shaking of wise heads, and mutterings
about storms that were brewing somewhere away, and
of high tides ; and the safety of the banks had to be
seen to. Birds were the weather omens, for good or
for ill, and our folks were rarely or never misled in
heeding them. Light-heartedness was a rare quality
on our shores, heredity and the force of circumstances made the natives grave and solemn of de-
meanour.

If they detested one thing more than another, it
was having a case of any kind brought before ' the
justices.' Matters were usually fought out between
man and man, and the system worked well. Tenacious enough they were, each one for what he knew
belonged to him, such as fur, fin, and feather ; also
concerning their various boundary marks. As a rule,
all made common cause, helping one another when
danger threatened or help was needed. News spread
quickly over the flats, for the graziers had good
horses ; no better of their sort were to be had.

One marshland farm I often visited stood by itself

in the marsh, full five miles away from any other dwelling.    You reached it by a marshland road, by no means an easy one to travel over, or you could branch off down to it from the upland hard main road, that overlooked nothing but marsh and water for miles upon miles, by a narrow track just wide enough for one waggon to go up or down it.    The farm-house and buildings, with one or two substantial cottages for the farm ' lookers,' as they were called, stood on the end of a slope, one of those long strips of solid ground that here and there rise gently up from the swamp-lands and join on to the uplands.

The main house was a large square brick build-ing, of two stories, having  huge  chimney-stacks ; solid it was throughout.    The upper windows were lead lights, so were most of the lower ones, on strong iron casements ; all the woodwork was of oak, dark with age, but sound as a bell.    The windows of the large parlour and living room, where the master and his family enjoyed the small leisure they could allow themselves,  had  very  strong  window-sashes  with numerous small squares of glass, all being protected by inside as well as outside shutters.    These windows opened on to the large walled-in garden, where figs, peaches, and nectarines ripened to perfection after the burning marshland summers.    In the latter part of autumn, just before the first frosts came, the trees were covered up for the winter by stout reed screens ; great sedge litter covered their roots. When the winter had fairly set in, and the marshes were covered with snow, the farm was a beacon by night and a landmark by day for those who were compelled to travel over them.

A noted stock-breeder was the master. Horned cattle and sheep he turned his attention to, and with great success. Where beasts range through sedges and willow-scrub thickets, and are out night and day from the late spring till early autumn, they are apt to get into very primitive ways, and the 'lookers' required all the assistance their strong intelligent dogs could give them, as well as to use freely their leaping-poles. These were of ash-wood, and they had a circular piece at one end to prevent the pole from bending, when a leap over the wide dykes was taken. Some of these dykes the dogs—grey, rough-coated, bob-tailed sheep-dogs with great brown eyes—had to swim across. In gadfly time it was a fine sight to see a herd of cattle charging along, bellowing, tails up and heads down, their horns rattling against each other like great sticks. A fine sight, at least, if you happened to be in the adjoining marsh, with a broad dyke between you and that herd. Otherwise you did well to climb on to the top of a big pollard willow, if one was within reach.

I was exploring a dyke one day that was fringed with the very finest marsh tangle I had ever seen. There were high plumed grasses, great burdocks with prickly burs that fasten tenaciously on to your clothing, rankly growing to the height of a man's head. Water-docks with great leaves made rare cool cover for all the creatures that were about. I was in search of the rarer kinds of rails, but had to abandon the quest as I came on something that turned the current of my thoughts very effectually. One of our very active marsh cows had conducted her calf hither, as a refuge from the heat, and, by gently pressing her broad muzzle over the young thing, had given it to understand that it should lie down there. Down lay the calf, out of sight under the cool large leaves, and the cow, fully convinced that her offspring was comfortable and in safe quarters, had gone a few yards further to feed on some delicate grass-tops.

Somehow or other, I managed to stumble on that hidden calf, which forthwith jumped up, blaring out its terror. There was a bellow, and a crash ensued ; then some large red and white object flashed through the tangle before my eyes. A very confused idea of the scene remained in my memory : a kind of cow earthquake—a bellowing, pounding and tearing up of turf, with loud snortings. For a moment I was spellbound ; then a strong arm hurled me on one side, whilst a voice roared out ' Git back ! ' That I did, and quickly too. Domestic animals, when left to their own devices in large extents of pasture, use instinctively the same methods for the protection of their young or their own welfare that they would in a wild state.

Sometimes, where the grazier's grounds were very large, as many as three, or even four, ' lookers ' would be employed to watch over them. Their office was no sinecure, for by turns these men acted as farm bailiffs, keepers, stock-tenders, fishers, boatmen, and wildfowlers, as the varying seasons demanded. Broad-shouldered, deep-chested men they were as a rule, of dark complexion and determined mien ; a weakly ' looker ' would have been worse than useless. I said they acted as keepers ; it was not because game needed looking after ; that kept to the rising ground miles away. But in dry seasons rare covers of partridges would come whirring down into the marshes for emmets, as the ants were always called, and the grasshoppers that swarmed in every direction. They did not stay till the shooting season, though ; after they had enjoyed a plentiful course of emmets and

' jumpers,' the birds whirred back again whence they came. A 'sou'-wester' covered the looker's head, one like that of a coalheaver, but smaller ; fishermen's boots reached up to the thigh, and a good guernsey down over the boot tops when these were drawn up.

Very anxious times those were for our graziers when high tides might be expected, especially in the night-time. The banks and sluices had all to be carefully inspected by the ' lookers,' each farmer having to keep in good condition the banks that protected his own grazing grounds. Where cattle can feed, hares will be found, rabbits also, in the drier parts of marshes especially. Huge warrens are the links or dunes in some counties, having a lagoon or fleet on the marsh side, and the open sea, or a wide creek where the salt water runs up, on the other. Unless the hares and rabbits can be confined to such places, they are killed as quickly as they are met with, and necessarily so, for the banks, or walls where the sluices are, protect thousands of cattle and their owners. A rabbit-burrow or two in a bank would effect more harm than could be remedied for years afterwards.

When the mushroom season was in, one man I knew well gave some of the 'lookers' plenty of exercise. It was a curious sight to see three of these run, then leap up in the air on their poles, as they came upon a dyke in pursuit of my agile friend the trespasser. This was the style of thing you heard: 'I say, "Flighter," look here! You'll cum onest too often, an' you knows it.' Or, 'You needn't take that 'ere pole by the middle; I ain't fool enough fur you tu ketch me flyin' over, where you've got out o' my ma'shes; but you'll git it; you've been out o' bounds agin.'

'I knows I hev', but I ain't got all the musherooms, you didn't give me time to fill this 'ere baskit. Now jest you go anuther track when I cums.'

'Sha'n't do nuthin' o' the sort. Dooty is dooty; an' ef you an' me does hev' a little ager medicine when I ain't on dooty, 'taint tu say as you are tu cum here musheroomin'; an' you wun't. I'll leather ye, or you'll leather me, if we does fall foul o' one anuther, when I'm on dooty.'

'I sha'n't be asleep when ye takes that 'ere job on; but I reckins I may git my musherooms out o' somebody else's ma'sh to-day, old Srimper!'

The afterglow of evening is on the marshes and lights up the waters of a winding lagoon which is fringed with reeds. The reed tassels look like dark feathers in the foreground; behind them the sky is a soft rose colour, streaked with bands of richest purple.

One great pole in the mid distance, which serves as a
guide for those at sea, towers up, clear from all.   All
is quiet; only a few bats skim over the reeds, just
touching the water now and again with their leathery
wings, and making small circles of light in the shadowy
portions of it.   A pair of marsh owls skim over, shoot
up in the air, dart over the reeds, and vanish again
with hawk-like speed.

From out of one of the dykes that lead into the
long watercourse a fowling punt glides like a shadow.
'Flappers' are about, and this is their time for coming out of the reeds on to the water.   Two young
fowlers are in the punt; one is in front with the long
gun, the other paddles her noiselessly along, the
ripples in her wake showing like dull red waves of
colour.   The punt glides on, and, with the exception
of a coot that clanks now and again, or the croaking
of a rail as he slips into the reeds and sedges, all is
still.   A few owls on the hunt sweep along over the
water, and into the marsh and back again; but as yet
no ducks show themselves, so the punt steals on.   Just
before it reaches a bend, the soft spattering sound of
ducks nozzling in soppy weed is faintly heard.   Very
gently the punt is brought over to the other side, into
deep shadow; then her head is turned towards the
point whence the sound comes, and there are visible

a nice lot of flappers with a few old ducks, all busily feeding.

A report rings out which shakes the reeds, and most of the fowl lie dead on the water. Swiftly the punt is paddled to get the cripples, and all are gathered save one which flutters to a spot which appears from the punt to offer firm footing. In his eagerness the fowler forgets to try the bottom with his pole before he springs towards his bird. With horror, his companion sees him go down, not in water but in quake bog, where he is smothered, buried out of sight in a few instants. As in a dream, the friend got home to tell the story. The body was recovered, but to this day the spot is shunned as though accursed.

Ay, more than these sad memories haunt one. How well I remember that bright summer morning when a party of our young fellows left the village for a swim. The tide was up; the bright water glittered in the sunlight, and the larks sang loudly all over the flats—the air seemed full of them. Like so many water spaniels, the boys took the water, and one, bolder than the rest, made for mid-channel, breasting it bravely. Sharp and loud over the water comes a cry for help; cramp has seized the brave swimmer. At their utmost speed his companions make for him. 'Tear through it, boys, or he's done for!'

A boat is put out from the side of a ship at anchor. 'Give way, lads, smart!' and the boat shoots through the water at racing pace, the oars going like clockwork. 'Bend to it, lads; he's down for the second time!' No need to urge them, they are pulling for life.

The swimmers know that the boat will reach him first, but, with their hearts in their mouths, on they swim.

He shows for the last time. 'Pull, lads, we shall have him! one more spurt for the love of God!' Too late. 'Well rowed, lads! Try the grapples!' shouts one of the swimmers. But the tide has carried him no one knows where now. And the seaman's practised eye telling him the swimmers are well-nigh exhausted, he bids them get into the boat quick. On reaching the shore they put on their clothes with heavy hearts, and ask who will carry the tidings to his mother. The ringleader in all their mad freaks and pranks—the boldest and most venturesome of the party—is asked to do it; but, with hot tears in his eyes, he refuses. She got the news quickly enough, poor soul, for a little lad who had been minding our clothes ran home in a fit of terror, telling everyone he met on his way up the street that poor Ned was drowned. 'Ah,' says one of the old fishermen, 'I knows where the

tide 'ull take him to when it turns—to the big sluice
it 'ull take him, see if it don't.'

We shudder as we think of the place, with its
massive piles and gates covered with sea-weeds and
tangle ; a place you would not want to look at twice,
when the tide was out. Imagine a gully twelve feet
deep, reaching to the base of the sea-wall, the sides
of it for a long distance lined with great piles—trees
pointed at one end and driven down into the ooze
thirty feet or more. Even with these you could see
the rush and wear of the tide : a grim place to look
at ! likely to give you the nightmare. At low-water
you could see great eels twisting about, and crabs,
those useful but ferocious scavengers, scuttling about
sideways, in search of food. A gruesome place,
shunned by all of us lads, especially at night, bold
though we were ; for we knew what had been found
there more than once.

About two hours after the tide had gone down the
heavy tread of fishermen in their great boots rang on
the pavement. They had found him in the big sluice.
From that time we avoided the spot more than ever.
To this day memory brings the picture of it all often
vividly before my eyes.

A curious thing happened when I was a boy, which
I have never seen mentioned anywhere. Hundreds

of French partridges came to our shore from seawards, and there dropped and lay exhausted. Some of our folks filled baskets full with them before they could recover sufficiently to get inland. They were in prime condition. That was in the afternoon ; the next day not a single bird of that species could be found. Some way or other they had certainly made a mistake in their reckonings. A calm day it was, too, without any wind. That summer was a glorious one ; many of the migrating birds stayed very late. Some bitterns were shot—the 'yaller French herns'—so called because they were rather common on the Essex coast and some portions of the Kentish flats at the time we were at war with France, or, as our folks said, 'that year we fit Old Boney.'

Spots such as that called 'the Marsh Fleet' are fast vanishing day by day. It lay in Kent, close to the Essex shore.

As we near the Fleet, or lagoon, for such it really is, the sun floods the whole extent of marsh and distant shore in a soft golden light. The cattle and the sheep look almost twice their natural size as they stand or lie half hidden in the long lush vegetation. Sails of vessels show in fine contrast to the green of the flats—some a flash of warm yellow, others gleaming red in the sun. The craft make their way slowly

along, shadow after shadow falling on their broad
sails as they pass or near each other. Most of them
are barges making for the mouth of the Thames or
the Medway. This lagoon is fringed with a belt of
high reeds and rushes for some distance out. By
careful management it is possible to squeeze into
them without getting fast in the mud ; but you must
step on the matted roots—break through them, and
there is no saying where you will go.

We have only made our way a few yards when
the birds let us know that we are trespassing. Reed
sparrows, or wrens, as they are called, chide and
chatter, running up the reed stems in a most dis-
tracted manner, for close to my face are one or two
of their nests. How deep that mud is we have no
means of knowing. Once I made a practical guess at
it to my sorrow. Decayed water-plants have left their
remains there year after year ; matter has been deposited from the water itself—all forming a light
flooring of unknown depth. The reed-cutters will
drop their long ash poles which they use to work
their punts with, and show you how far down they
will go with a simple pressure of the hands.

But changes have come over our flats, and time
has made a difference to all our longshore dwellers.
One place, once a celebrated resort of wildfowl, is

now a fruit orchard ; and a part of the shore frequented by sanderlings and dotterels, each in their season, is now covered with fine houses, forming a marine parade.

And so our longshore shooter's occupation is gone, and, if he would earn a living, he must often take to very different work, to wit, the drainage of his much-loved marshes.

PRINTED BY
SPOTTISWOODE AND CO., NEW-STREET SQUARE
LONDON

# SMITH, ELDER, & CO.'S PUBLICATIONS.

**THE WHITE COMPANY.** By A. Conan Doyle, Author of 'Micah Clarke' &c. Eighth Edition. Crown 8vo. 6s.

**THE NEW RECTOR.** By Stanley J. Weyman, Author of 'The House of the Wolf' &c. Crown 8vo. 6s.

**THE HISTORY OF DAVID GRIEVE.** By Mrs. Humphry Ward, Author of 'Robert Elsmere' &c. Popular Edition. Crown 8vo. 6s.

**ROBERT ELSMERE.** By Mrs. Humphry Ward, Author of 'Miss Bretherton' &c. Cheap Edition, crown 8vo. limp cloth, 2s. 6d.
\*₌\* Also the POPULAR EDITION, 1 vol. crown 8vo. 6s.; and the Cabinet Edition, 2 vols. small 8vo. 12s.

**THE GAMEKEEPER AT HOME;** or, Sketches of Natural History, Poaching, and Rural Life. By Richard Jefferies. New Edition, with all the Illustrations of the former edition. Crown 8vo. 5s.

*By the same Author.*

**WILD LIFE IN A SOUTHERN COUNTY.** New Edition. Cr. 8vo. 6s.

**HODGE AND HIS MASTERS.** New Edition. Crown 8vo. 7s. 6d.

**THE AMATEUR POACHER.** New Edition. Crown 8vo. 5s.

**ROUND ABOUT A GREAT ESTATE.** Crown 8vo. 5s.

**WOODLAND, MOOR, AND STREAM;** being the Notes of a Naturalist. Edited by J. A. Owen. Second Edition. Crown 8vo. 5s.

**FALLING IN LOVE;** with other Essays treating of some more Exact Sciences. By Grant Allen. Cheap Popular Edition. Fcp. 8vo. limp green cloth, or cloth boards, gilt top, 2s. 6d.

**A BRIDE FROM THE BUSH.** By E. W. Hornung. Crown 8vo. limp red cloth, 2s. 6d.

**SIX MONTHS IN THE RANKS;** or, the Gentleman Private. Crown 8vo. limp red cloth, 2s. 6d.

**CHARLES FRANKLYN OF THE CAMEL CORPS.** By Hasmbib. Crown 8vo. 6s.

**HOLIDAY PAPERS.** Second Series. By the Rev. Harry Jones, Author of 'East and West London' &c. Crown 8vo. 6s.

**JESS.** By H. Rider Haggard, Author of 'King Solomon's Mines' &c. Crown 8vo. limp red cloth, 2s. 6d.

**VICE VERSA;** or, a Lesson to Fathers. By F. Anstey. Crown 8vo. limp red cloth, 2s. 6d.

*By the same Author.*

**A FALLEN IDOL.** Crown 8vo. 6s. Cheap Edition, crown 8vo. limp red cloth. 2s. 6d.

**THE GIANT'S ROBE.** Crown 8vo. 6s. Cheap Edition crown 8vo. limp red cloth, 2s. 6d.

**THE PARIAH.** Crown 8vo. 6s. Cheap Edition, crown 8vo. limp red cloth, 2s. 6d.

**THE TALKING HORSE,** and other Tales. Second Edition, crown 8vo. 6s.

**MORE T LEAVES;** a Collection of Pieces for Public Reading. By Edward F. Turner, Author of 'T Leaves,' 'Tantler's Sister,' &c. Cr. 8vo. 4s. 6d.

*By the same Author.*

**T LEAVES;** a Collection of Pieces for Public Reading. Sixth Edition. Crown 8vo. 3s. 6d.

**TANTLER'S SISTER; AND OTHER UNTRUTHFUL STORIES:** being a Collection of Pieces written for Public Reading. Third Edition. Crown 8vo. 3s. 6d.

London: SMITH, ELDER, & CO., 15 Waterloo Place.

# WORKS BY F. ANSTEY.

SECOND EDITION. Crown 8vo. 6s.

## THE TALKING HORSE;
### AND OTHER TALES.

**From THE SATURDAY REVIEW.**—'A capital set of stories, thoroughly clever and witty, often pathetic, and always humorous.'

**From THE ATHENÆUM.**—'The grimmest of mortals, in his most surly mood, could hardly resist the fun of "The Talking Horse."'

POPULAR EDITION. Crown 8vo. 6s.
CHEAP EDITION. Crown 8vo. limp red cloth, 2s. 6d.

## THE GIANT'S ROBE.

**From THE PALL MALL GAZETTE.** —'The main interest of the book, which is very strong indeed, begins when Vincent returns, when Harold Caffyn discovers the secret, when every page threatens to bring down doom on the head of the miserable Mark. Will he confess? Will he drown himself? Will Vincent denounce him? Will Caffyn inform on him? Will his wife abandon him?—we ask eagerly as we read and cannot cease reading till the puzzle is solved in a series of exciting situations.'

POPULAR EDITION. Crown 8vo. 6s.
CHEAP EDITION. Crown 8vo. limp red cloth, 2s. 6d.

## THE PARIAH.

**From THE SATURDAY REVIEW.** - 'In "The Pariah" we are more than ever struck by the sharp intuitive perception and the satirical balancing of judgment which makes the author's writings such extremely entertaining reading. There is not a dull page—we might say, not a dull sentence—in it. . . . The girls are delightfully drawn, especially the bewitching Margot and the chi'dish Lettice. Nothing that polish and finish, cleverness, humour, wit, and sarcasm can give is left out.'

CHEAP EDITION. Crown 8vo. limp red cloth, 2s. 6d.

## VICE VERSÂ;
### OR, A LESSON TO FATHERS.

**From THE SATURDAY REVIEW.**—'If ever there was a book made up from beginning to end of laughter, and yet not a comic book, or a "merry" book, or a book of jokes, or a book of pictures, or a jest book, or a tomfool book, but a perfectly sober and serious book, in the reading of which a sober man may laugh without shame from beginning to end, it is the new book called "Vice Versâ ; or, a Lesson to Fathers." . . . We close the book, recommending it very earnestly to all fathers in the first instance, and their sons, nephews, uncles, and male cousins next.'

CHEAP EDITION. Crown 8vo. limp red cloth, 2s. 6d.

## A FALLEN IDOL.

**From THE TIMES.**—'Mr. Anstey's new story will delight the multitudinous public that laughed over "Vice Versâ." . . . The boy who brings the accursed image to Champion's house, Mr. Bales, the artist's factotum. and above all Mr. Yarker, the ex-butler who has turned policeman, are figures whom it is as pleasant to meet as it is impossible to forget.'

London : SMITH, ELDER, & CO., 15 Waterloo Place.

# ILLUSTRATED EDITIONS
## OF
# POPULAR WORKS.

*Handsomely bound in cloth gilt, each volume containing Four Illustrations. Crown 8vo. 3s. 6d. each.*

THE SMALL HOUSE AT ALLING-TON. By ANTHONY TROLLOPE.

FRAMLEY PARSONAGE. By ANTHONY TROLLOPE.

THE CLAVERINGS. By ANTHONY TROLLOPE.

TRANSFORMATION: a Romance. By NATHANIEL HAWTHORNE.

DOMESTIC STORIES. By the Author of 'John Halifax, Gentleman.'

THE MOORS AND THE FENS. By Mrs. J. H. RIDDELL.

WITHIN THE PRECINCTS. By Mrs. OLIPHANT.

CARITÀ. By Mrs. OLIPHANT.

FOR PERCIVAL. By MARGARET VELEY.

NO NEW THING. By W. E. NORRIS.

LOVE THE DEBT. By RICHARD ASHE KING ('Basil').

WIVES AND DAUGHTERS. By Mrs. GASKELL.

NORTH AND SOUTH. By Mrs. GASKELL.

SYLVIA'S LOVERS. By Mrs. GASKELL.

CRANFORD, and other Stories. By Mrs. GASKELL.

MARY BARTON, and other Stories. By Mrs. GASKELL.

RUTH; THE GREY WOMAN, and other Stories. By Mrs. GASKELL.

LIZZIE LEIGH; A DARK NIGHT'S WORK and other Stories. By Mrs GASKELL.

# POPULAR NOVELS.

*Each Work complete in One Volume, Crown 8vo. price Six Shillings*

THE HISTORY OF DAVID GRIEVE. By Mrs. HUMPHRY WARD, Author of 'Robert Elsmere' '&c.

THE WHITE COMPANY. By A. CONAN DOYLE, Author of 'Micah Clarke.'

THE NEW RECTOR. By STANLEY J. WEYMAN, Author of 'The House of the Wolf' &c.

NEW GRUB STREET. By GEORGE GISSING.

EIGHT DAYS. By R. E. FORREST, Author of 'The Touchstone of Peril.'

A DRAUGHT OF LETHE. By ROY TELLET, Author of 'The Outcasts' &c.

THE RAJAH'S HEIR. By a NEW WRITER.

THE PARIAH. By F. ANSTEY, Author of 'Vice Versâ' &c

THYRZA. By GEORGE GISSING, Author of 'Demos' &c.

THE NETHER WORLD. By GEO. GISSING, Author of 'Demos' &c.

ROBERT ELSMERE. By Mrs. HUMPHRY WARD, Author of 'Miss Bretherton' &c.

RICHARD CABLE: the Lightship-man. By the Author of 'Mehalah,' 'John Herring,' 'Court Royal,' &c.

THE GAVEROCKS. By the Author of 'Mehalah,' 'John Herring,' 'Court Royal,' &c.

A FALLEN IDOL. By F. ANSTEY, Author of 'Vice Versâ' &c.

THE GIANT'S ROBE. By F. ANSTEY, Author of 'Vice Versâ' &c.

DEMOS: a Story of Socialist Life in England. By GEORGE GISSING Author of 'Thyrza' &c.

LLANALY REEFS. By Lady VERNEY, Author of 'Stone Edge' &c.

LETTICE LISLE. By Lady VERNEY. With 3 Illustrations.

OLD KENSINGTON. By Miss THACKERAY.

THE VILLAGE ON THE CLIFF. By Miss THACKERAY.

FIVE OLD FRIENDS AND A YOUNG PRINCE. By Miss THACKERAY.

TO ESTHER, and other Sketches. By Miss THACKERAY.

BLUEBEARD'S KEYS, and other Stories. By Miss THACKERAY.

THE STORY OF ELIZABETH; TWO HOURS: FROM AN IS-LAND. By Miss THACKERAY.

TOILERS AND SPINSTERS, and other Essays. By Miss THACKERAY.

MISS ANGEL: Fulham Lawn. By Miss THACKERAY

MISS WILLIAMSON'S DIVAGA-TIONS. By Miss THACKERAY.

MRS. DYMOND. By Miss THACKERAY.

London: SMITH, ELDER, & CO., 15 Waterloo Place.

www.ingramcontent.com/pod-product-compliance
Lightning Source LLC
Chambersburg PA
CBHW021942220326
41599CB00013BA/1487